MANAGEMENT OF GEOLOGICAL DATABASES

COMPUTER METHODS IN THE GEOSCIENCES

Daniel F. Merriam, Series Editor

Volumes in the series published by Pergamon:

Geological Problem Solving with Lotus 1-2-3 for Exploration and Mining Geology: G.S. Koch Jr. (with program on diskette)
Exploration with a Computer: Geoscience Data Analysis Applications: W.R. Green
Contouring: A Guide to the Analysis and Display of Spatial Data: D.F. Watson (with program on diskette)

*Volumes published by Van Nostrand Reinhold Co. Inc.:

Computer Applications in Petroleum Geology: J.E. Robinson
Graphic Display of Two-and Three-Dimensional Markov Computer Models in Geology: C. Lin and J. W. Harbaugh
Image Processing of Geological Data: A.G. Fabbri
Contouring Geologic Surfaces with the Computer: T.A. Jones, D.E. Hamilton, and C.R. Johnson
Exploration-Geochemical Data Analysis with the IBM PC: G.S. Koch, Jr. (with programs on diskettes)
Geostatistics and Petroleum Geology: M.E. Hohn
Simulating Clastic Sedimentation: D.M. Tetzlaff and J.W. Harbaugh

*Orders to: Van Nostrand Reinhold Co. Inc, 7625, Empire Drive, Florence, KY 41042, USA.

Related Pergamon Publications

Books

GAAL & MERRIAM (Editors): Computer Applications in Resource Estimation: Prediction and Assessment for Metals and Petroleum

HANLEY & MERRIAM (Editors): Microcomputer Applications in Geology II

Journals

Computers & Geosciences

Computer Languages

Information Processing & Management

International Journal of Rock Mechanics and Mining Sciences (& Geomechanics Abstracts)

Minerals and Engineering

Full details of all Pergamon publications/free specimen copy of any Pergamon journal available on request from your nearest Pergamon office.

MANAGEMENT OF GEOLOGICAL DATABASES

Joseph Frizado

Chairman, Department of Geology,
Bowling Green State University,
Bowling Green, Ohio, USA

PERGAMON PRESS

OXFORD · NEW YORK · SEOUL · TOKYO

U.K.	Pergamon Press Ltd, Headington Hill Hall, Oxford OX3 0BW, England
U.S.A.	Pergamon Press, Inc, 660 White Plains Road, Tarrytown, New York 10591-5153, U.S.A.
KOREA	Pergamon Press Korea, KPO Box 315, Seoul 110-603, Korea
JAPAN	Pergamon Press Japan, Tsunashima Building Annex, 3-20-12 Yushima, Bunkyo-ku, Tokyo 113, Japan

---

First edition 1992

Library of Congress Cataloging in Publication Data
A catalogue record for this book is available from the Library of Congress.

British Library Cataloguing in Publication Data
A catalogue record for this book is available from the British Library.

ISBN 0 08 037951 6

Printed in Great Britain by BPCC Wheatons Ltd., Exeter

Contents

Series Editor's Foreword

Database management and microcomputers - two magic words in todays highly sophisticated research environment. This book with those words in the title is the result of a workshop held in 1987 in Kuwait for a group of scientists from all over the world to learn of the latest developments in these areas. This material now is made available to you. Database management, of course, is necessary to handle the vast amounts of data available with automated data-acquisition. The microcomputer, or PC, is powerful enough to manipulate and analyze just about anything necessary to carry on meaningful research. Combine the two - a database and the PC - and you have the tools with which to work that were unimagined just a few years ago.

Here then is an expose of these subjects with emphasis on "hardrock geology" compiled by Joe Frizado and colleagues. The book takes the beginner through the procedures step-by-step with explanations on the accompanying computer programs used to analyze data in a large database. The seven chapters include the Introduction; a DBMS (database management system); an explanation of the programs; use of spreadsheets; interpreting the data; stream-sediment database for mineral exploration for mineral deposits from igneous rocks.

From the chapter contents it should be obvious that the first part of the book is an introduction and explanation to DBMS and that the second part contains the applications and examples. The intent of the book according to Frizado is to introduce and to facilitate the "...use of microcomputer database management systems (DBMS) on geological data." The book is well written and easy to follow. The reader should have no trouble in fulfilling his or her objective of learning the basics of the microcomputer DBMS.

This book is yet another in a long series of books to bring to the researcher and practitioner techniques and applications on the latest quantitative aspects of geology to enhance their ability to do a better job quicker in this fast-moving "Information Age."

D. F. Merriam

Preface

This volume has its origins in work performed in International Geological Correlation Program Project #239. The primary goal of the project was to promote the use of a worldwide database on igneous rocks (IGBA). As the project developed, we widened this goal to include facilitating the use of microcomputer database management systems (DBMS) on geological data.

As part of this effort, with support from UNESCO, the International Union of Geological Sciences, United States National Science Foundation, Kuwait University, and Bowling Green State University, we offered a workshop for geoscientists in the use of microcomputer-based database management systems, in September 1987, in Kuwait. During a two week time span, with a mixture of lecture, reading and hands-on experience, more than 50 scientists from 37 countries were introduced to the basics of DBMS systems, and learned about the geological applications of such systems in academic settings, in exploration for mineral resources and in research.

Given enough time and documentation, anyone can learn how to use a microcomputer program. The language and techniques used in database management may be overwhelming and discouraging. A geologist introducing himself or herself to the subject without external assistance of the sort offered here, easily may fail to realize that even rudimentary knowledge of DBMS techniques may be sufficient for many useful applications in earth science. For this reason, the Kuwait workshop included only enough basic instruction in two microcomputer DBMS programs to get the participants through the steep part of the learning curve. The second portion of the course was spent studying the application of DBMS programs to actual geological problems and data. By interspersing lectures devoted to explaining operating techniques with practical sessions in which participants themselves extracted and organized data, we attempted to keep the interest of the participants alive and in fact stimulated many to begin planning how they could apply the same techniques to their own projects.

This intimate admixture of explanation and application is apparent also in the organization of our book, which is divided into two sections. The initial portion is a guide to the basics of selecting a database management system as well as an introduction to two widely used DBMS programs, PC-File and dBASE. Each of the programming chapters emphasizes a hands-on approach. As you read a section, it will be helpful to have the relevant program running so that you can see the results as you proceed. The accompanying data disk* contains copies of the data files for completing the exercises described in this volume. (Of course you cannot actually do the exercises unless you have the requisite software. You may not have access to either database management system used in our examples, but you can do the exercises in these chapters with almost any other microcomputer DBMS.)

The fourth chapter of the book introduces the reader to spreadsheets and rudimentary graphics. Having extracted data from a database, a spreadsheet is a useful vehicle for manipulating data extracted from any database. You soon will discover that it also can be used itself as an elementary information management system.

The last three chapters of the book describe a series of applications of database management to geological problem solving, so that you can begin to see how to use the techniques introduced in part one. These later chapters demonstrate the convenience and power of database management techniques in a variety of different geological contexts. The second half of the book also contains a few exercises that can be used, as necessary, to develop further your data management expertise.

As microcomputers have become a standard tool of geologists, the ability to store vast amounts of information has become commonplace.

* available from Joseph Frizado

Without the aid of data management programs, it is easy to be overcome by the wealth of information that can be collected. Our book is designed to encourage you to take the first steps towards efficient geological database management. It should allow you get started quickly in building your expertise in the use of database management programs. It is not meant to be a standard reference. Rather, our hope is that it will provide an easy and efficient path to a "user's eye" understanding of the general structure of database management programs, and how they can be used in your own work in geology.

Joseph Frizado

CHAPTER 1

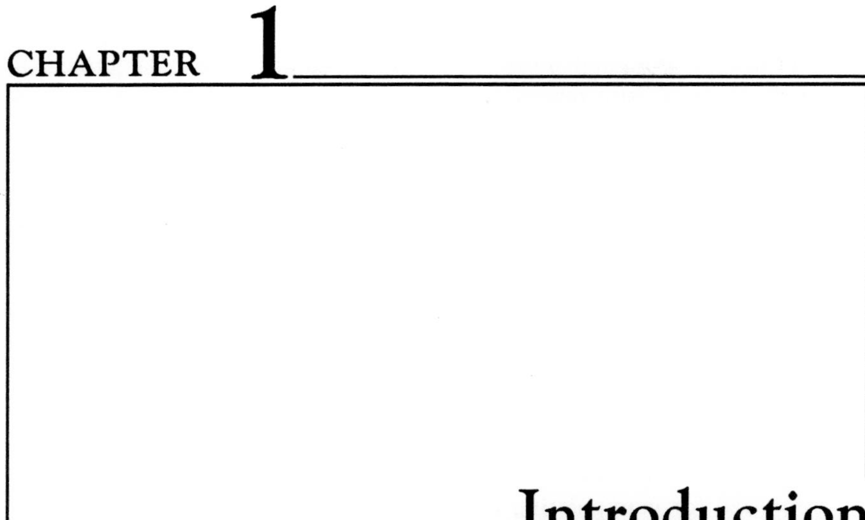

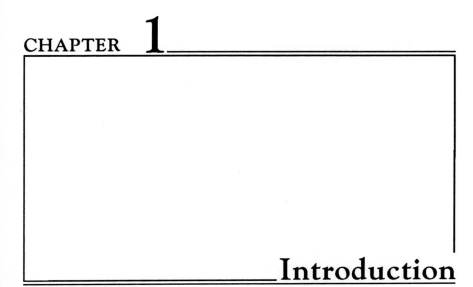

Introduction

Joseph Frizado
Dept. of Geology
Bowling Green State University
Bowling Green, Ohio USA

1.1 THE DECISION-MAKING PROCESS

The management and manipulation of data is one of the essential processes of science. The ability of computers to perform repetitive tasks quickly and without error makes it a natural tool for managing large amounts of information. However, it is up to the user to determine what data should be stored in machine readable form, how it should be stored and what programs/machines will be used to manage the data. The large number of database management programs for mainframe computers, minicomputers and microcomputers offers the geologist many choices, but the basic principles of different types of systems can be described in similar terms. Hence, the decision making process for petrologic database management can be generalized into a series of steps:

(1) Is the database that I am going to use large enough or complex enough to require computer assistance? In all instances, this a personal question. If the database that you are to use on a particular project is small, there is little reason to utilize a computer. Also, if you cannot acquire easily a computer for such work, a database management program would be of little use to you. This quickly becomes a cost/benefit calculation for the user. The pluses of speed and ability to perform complex searches in a large block of information must be weighed against learning new programs/languages, acquiring or gaining access to equipment, and other financial costs. We must remember that it is difficult to kill a gnat with an elephant gun and definitely is not worth the trouble.

(2) What is the nature of the data to go into the database? If the base that I am going to use contains primarily character data and little numeric data, then I will want to utilize a database management system that is optimized for manipulating characters and searching character strings for particular combinations of characters. On the other hand, petrological databases tend to be quantitative and the ability to manipulate and search numeric fields usually is much more important. How many data elements will be recorded for each entity? What is the maximum size for each possible entry? Both of these questions lead to the next step.

(3) What is the approximate size of the database? The size of the database can affect our decision in two important characteristics: memory/disk capacity and speed. Most computer based management systems will take an entire description of an entity (called a record) into random access memory (RAM). If the record is large, then smaller computers may not have enough RAM to perform the required operations. This also is true in data storage materials. If a record will occupy a given size, how many records will be contained on the usual storage material for that computer (be it tape, floppy disk, hard disk, etc.). Some computer programs cannot work over several disks or several distinct blocks of data from different sources. Speed becomes a consideration at this point also. If you are dealing with a large database, then the speed of your computer and of your program become important. On a small database most programs work quickly, but on a larger data set inefficiencies in programming

and a slower computer can cause tremendous time delays which can be costly.

(4) Am I constrained by current applications? If my organization has a computer that I have access to, this question may become the top priority in the decision making process. If my organization has an IBM PC AT and I can purchase a database management system for my work, I probably will be limited to those programs produced for this machine. If on the other hand, my budget includes the possibility of purchasing any type of machine I choose, then this question is less of a constraint. I may be limited further in my decision by my organization already having a particular database management system with a particular computer. At this point the question really should be rephrased into "Can I do my project using that system?" or "How can I do my project on that system in the most efficient manner?". Unfortunately, many of us can only react to someone else's choice of machine and software.

Once those questions are answered by a particular user, the choice of a particular database management system can be pursued. In each of the subsequent steps, the decision maker has to be aware of constraints placed on the decision by the previous steps. The next question to be answered is whether a turn-key application program is required, or should I write my own series of programs for data management. In some instances, where the data is highly specialized or the operations to be performed on the data are sufficiently specific, writing your own series of file and data management programs is probably the best route to take. However, if the database is large or if you wish to perform complex searches and wish to link to different databases, a commercial product becomes more appropriate.

1.2 TYPES OF DATABASE MANAGEMENT SYSTEMS

All database management systems use a particular vocabulary for describing the data within the system. A field is the smallest unit of data entered into the base. It is a particular value or string of letters and numbers that cannot be broken down into any smaller units without losing information. For example, a database may have a field for weight percent SiO_2 which would be filled with a numeric value between 0 and 100%. The field is identified by a field name, in this instance SIO2. The value for the weight percent SiO_2 of a particular specimen is placed in that field location in the database. A collection of fields that are related is grouped together into a record. A record is a collection of values within the fields that usually describe a single item or sample. For example a record could contain SiO_2, Al_2O_3, TiO_2, FeO, etc. of a single rock specimen. This record also could contain other information about the sample, such as location, sample number, and rockname. The record will contain all of the relevant information about that particular sample or group of samples. A database is made up of a series of records each of which contains a series of fields. In most databases, each record has the same set of fields and would be akin to a paper form to be filled out describing a sample. The fields are the entries on the paper form. In the usual parlance, a database file is a collection of records.

Turn-key or off-the-shelf database management programs can be divided into three general groups: flat-file, hierarchical, and relational. Each group of programs uses a different view of the data and its interrelationships within the database. Each has its pros and cons for a particular application.

Flat-file databases are used mostly in the simplest microcomputer database management programs (such as pfs:file, dbMaster, pc-File, etc.). A flat-file system looks at each record as being part of a single file. The records are entered sequentially. This is analogous to a card file system. Each record is a card in the box, each bit of information on the card is a field. Anytime a search of the data file is performed, all records are scanned for the particular data entry or condition, before being selected or rejected in the search. In this instance, the entire database almost always is processed in a search. Searching can be improved by indexing the file. An index file can be created by sequencing records with regard to the value of a particular field on each record and storing the location of each record in the index file. For example, a database can be indexed on SiO_2. The index file would have a series of values for SiO_2 and with each of those values, the record number or position of record(s) that have that SiO_2 value. Once this is accomplished, one can easily select samples within a particular range of SiO_2 contents by scanning the sorted index of values and noting the record locations of those records meeting the criteria. The speed of execution also can be improved by sorting the database. The position of records within the database can be based upon their position in the index. For example, the record with the highest value of SiO_2 can be placed as the first record in the database. The lowest SiO_2 content would be the last record in the base. Because all of the records with similar SiO_2 are located in approximately the same location, access time is shortened. However, a sorted database only improves the speed of the system if you are searching with respect to the sorted value. If your next search is based on Al_2O_3, then a sorted SiO_2 base would not improve your speed. The time it takes for the system to generate a sorted version of itself also must be included in the analysis. If you are only going to search the base on SiO_2 a few times, you need to compare the time to run the sort and perform faster searches versus just performing slower searches. As a rule of thumb, do not sort the database unless the sort key usually is part of your search key when you use the base. Indices on the other hand improve speed and do not take much time to create. You can have several indices on a particular file based on different fields. You then can use the Al_2O_3 index to select records, just as easily as using the SiO_2 index. If you do not envisage making particular repetitive searches on a database, a series of indices to facilitate such searches may be more trouble than they are worth. As your use of a particular search pattern increases, you should go to an index to speed up a flat-file database.

Hierarchical file databases are useful when there is a particular structure to the data (Clemmons, 1985). In such a structure, records are linked via a parent-daughter relationship. Records at the top of the database, sometimes termed root, can be the parents to several records on the next layer of the database. This creates an inverted tree-like structure as depicted in Figure 1.1. The actual records under each parent may be stored there physically, or a pointer to a record can be used. The hierarchical structure can be useful and fast in particular circumstances. For example as a search progresses to lower and lower levels, you need not search large sections of the database. Searches can leap to a particular level in the tree structure and search only at that level to select records for inclusion. To use this structure effectively, however the programmer needs to know the data and its interrelationships to perform efficient searches. The structure of the base and the selection of what level in the

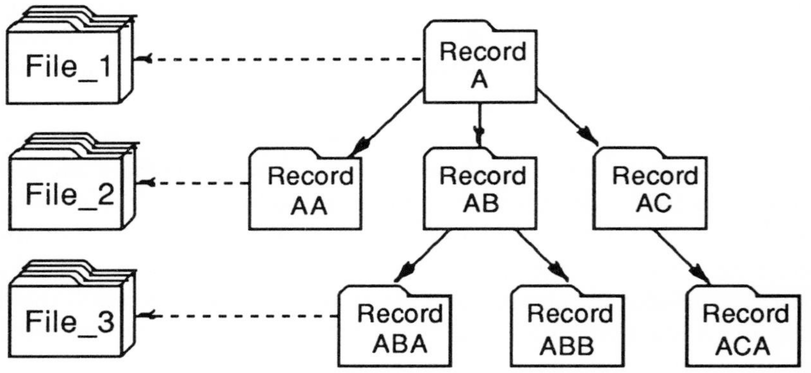

Figure 1.1 Hierarchical database structure.

system become important. For this reason, few novice users tend to use such programs.

Relational database management systems view the relationship between data records in a different manner than hierarchical managers. The relational data structure resembles a table or tabular format. Each row within a table corresponds to a series of fields for a particular sample and is therefore a record in this type of manager. Multiple files can be related by a general unique field contained in each of the linked data structures. This allows several databases to be linked via related fields, so that a single record in one database with particular characteristics also can delineate records in another database based upon a user-defined relationship between the two databases. A sequential file structure with fixed format records, each record is the same size and each record follows in a prescribed sequence, can be used in such database management systems, but more complex file structures are utilized in the more expensive commercial systems. At present, most of the more powerful database managers use a relational approach. General terms in discussing such a base refer to "tuples" which are equivalent to records or rows in the table form and "attributes" which are the fields or columns in the table (Clemmons, 1985). An example of such a format for a petrological database is given in Figure 1.2. Each sample number corresponds to a set of major element data for the sample. Each sample can have several entries in a trace element data table. Each bibliographic reference in a third table can have several samples in each reference. By using three different databases, but linked via these relationships, the user can perform operations on a smaller database, then retrieve the pertinent information for the linked bases as needed.

The choice between the different styles of database management systems often is made based upon price/performance characteristics. The hierarchical system and relational system tend to have a higher cost than flat-file systems. However, for small databases, flat-file system slowness is not a tremendous problem, and becomes the system of choice. Unfortunately, as the system price decreases, many other features may be dropped from consideration.

1.3 DESIGNING A DATABASE

Database design involves considerable planning before ever touching the keyboard. Once data has been entered into a database structure, the information can be moved into a different database, usually with difficulty and pain. It is preferable to plan ahead before creating the database in order to maximize your efforts. There are two points that you should have in mind during the design process: (1) how should the information be stored and (2) how will users ask for data retrievals? The actual design of a database can be achieved in a three step process: (1) data definition, (2) data refinement, and (3) establishing relationships between fields (Kahn ,1985; Jones, 1987).

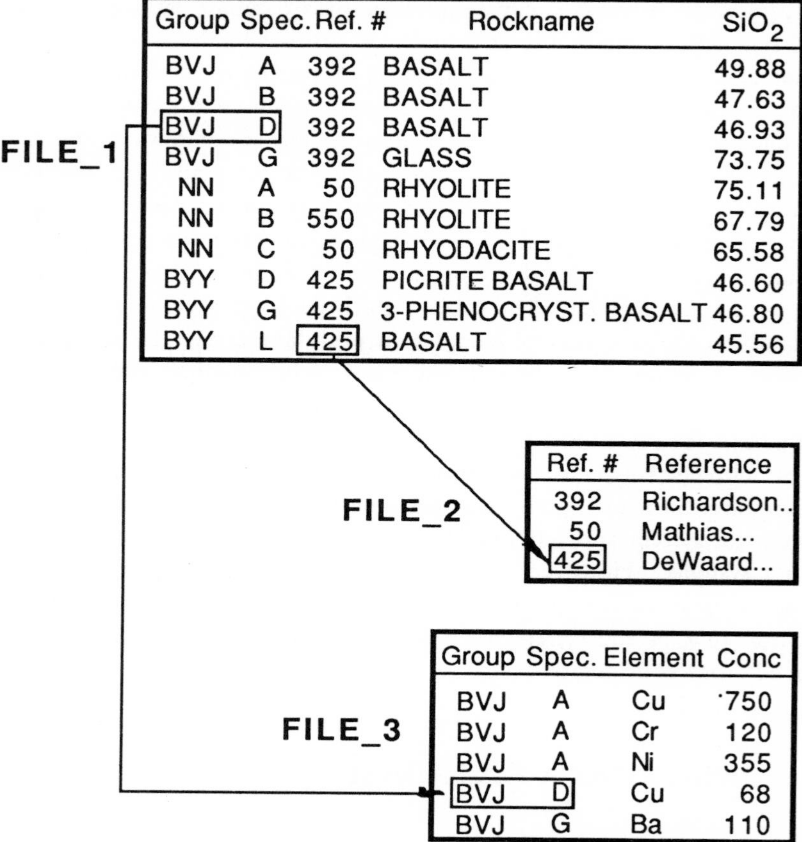

Figure 1.2 Relational database structure for petrological database. In this example three distinct databases are linked by particular fields: main database with major element data, trace element database, and bibliographic reference database.

The first step to take is to define what data will be going into the database. In our example case, we will be looking at data stored in a world-wide igneous rock database, IGBADAT (Chayes, 1985). This database has been compiled from the literature and includes locations, rock names, major element contents, trace element determinations, petrographic descriptions, mineral association data, ages (both stratigraphic and radiometric) and so on of igneous rocks. We also have bibliographic information because each sample in the database has come from a published article. All possible fields to be included within the database should be listed at this stage.

During the data refinement stage, the designer should look at the particulars of each field. In order to design a database for such geological information, we need to know what types of searches will be performed. In our example example, we will want a search based upon location, geological age, rockname and by numeric combinations of the major oxides. Of course, each of these fields will have to be in the main database. If we are using a flat-file system, all of the pertinent data will have to be entered. Once a complete list of what you wish to have in the database is compiled, the order of entry should be developed so that similar items are grouped together.

In a relational system, the database creator also must decide what information should go into what database. In our example, this can be demonstrated with the bibliographic information. Each article has several samples within the dataset. We have two options: (1) include in each sample record the complete bibliographic data, or (2) link a reference database to the main database so that each entry in the main database points to the correct reference in the reference database. The second option saves space by not having to duplicate data for all of the records from a single literature source.

As an additional task under data refinement, the designer also should begin to determine the parameters of each field. If a field is to contain character data, what is the longest string of characters to be contained there? Many databases will allocate the full amount of space designated for a character field. During the creation of the database, the designer will have to identify how many characters can be contained within a defined field. If this number is set too high, much wasted space can occur. If set too low, data will be lost. This also is true for setting the number of decimal places and the type of the numeric field. One method of conserving space within a database is the use of codes for particular circumstances. In our example of using major elements from IGBADAT (Chayes, 1985), we would be concerned with how the data was obtained. Rather than include long character strings that would be repeated through a series of samples, a code for particular circumstances can be worked out

(Table 1.1). By creating such a code prior to construction of the databases, the designer can keep the length of each record to a minimum while maintaining the maximal information in each record.

The final stage before starting the construction of your database is to define the relationships between fields. This process allows you to identify what fields are the most important and possibly what fields can be dropped from the structure. This entire process hinges on what the users

Table 1.1 Status symbol codes used in IGBADAT

Specimen location by latitude and longitude, in sources reference

1D	NOT LISTED OR SHOWN
1B	TO NEAREST DEGREE ONLY
1C	TO NEAREST TENTH OF DEGREE ONLY

'Completeness' of essential oxide analysis

2A	COMPLETENESS INCOMPLETELY SPECIFIED IN SOURCE DESCRIPTION
2B	ANALYSIS NORMALIZED TO 100% IN SOURCE
2C	FE-OXIDE PARTITION NOT DETERMINED ON ANALYSED SPECIMEN
2D	TOTAL IRON ONLY. STORED AT FEO, FE2O3, OR FE
2E	TOTAL H2O NOT DIRECTLY DETERMINED
2F	H2O NOT PARTITIONED
2G	H2O+ IS LOSS ON IGNITION
2H	SAMPLE AIR DRIED, DESICCATED DURING ANALYSIS
2I	SOME ESSENTIAL OXIDE(S) NOT DETERMINED

Analytical procedures and methods

3A	RESULT AN AVERAGE FOR MULTIPLE ANALYSES OF SAME SPECIMEN
3K	RESULT AN AVERAGE OF ANALYSES OF 2 OR MORE SPECIMENS
3N	ANALYSED SPECIMEN IS A STANDARD OR REFERENCE MATERIAL
3B	COMPOSITE SAMPLE USED FOR ANALYSIS

Table 1.1 cont'd

3L	REPLICATE ANALYSIS OF SPECIMEN
3M	'CORRECTION' OF A PREVIOUSLY PUBLISHED ANALYSIS
3C	SOME ESSENTIAL OXIDE(S) NOT QUOTED TO 0.01%
3D	ALKALIS DETERMINED BY FLAME PHOTOMETER
3E	SOME ESSENTIAL OXIDE(S) DONE BY X-RAY FLUORESCENCE
3J	SOME ESSENTIAL OXIDE(S) DONE BY ATOMIC ABSORPTION
3F	SOME ESSENTIAL OXIDE(S) DONE BY ELECTRON PROBE
3G	SOME ESSENTIAL OXIDE(S) DONE BY NEUTRON ACTIVATION
3H	RADIATION OTHER THAN XRF, EPR, NAC, ATAB, TLFT USED IN ANALYSIS
3I	SOME TRACE ELEMENT(S) DETERMINED BY ARC SPECTROGRAPHY
Associated data recorded in source description	
4J	NO PETROGRAPHIC INFORMATION GIVEN IN SOURCE REFERENCE
4K	NO MINERALOGICAL INFORMATION GIVEN IN SOURCE REFERENCE
4A	PETROG. DESC. GENERALIZED; MAY NOT APPLY TO ANALYSED SPEC.
4B	MIN. ASSOC. GENERALIZED; MAY NOT APPLY TO ANALYSED SPEC.
4C	STRAT. AGE INFERRED; MAY NOT APPLY TO ANALYSED SPECIMEN
4D	PHYSICAL AGE INFERRED; NOT DETERMINED ON ANALYSED SPEC.
4E	QUANTITATIVE MODAL ANALYSIS OF CHEMICALLY ANALYSED SPEC.

of the databases wish to search for. By formulating some of the questions to be answered once the database is complete, the designer can obtain a better idea of what an optimal structure might be like. If there is interest in knowing what is the average TiO_2 content of basalts in the database, you will need to have a rock name field as well as a TiO_2 field. If at the same time, you can see no need to keep the mineral assemblage data, drop that item from the database. However, one word of warning, try to consider as many possible future uses of your database before you decide not to keep a particular item. As a novice, it is better to err on the side of having too large a database rather than too small. It always is easier to make a small database of a selected group of fields from a larger one. The reverse may not be true.

References

Chayes, F., 1985, IGBADAT:a world data base for igneous petrology: *Episodes*, vol. 8, no. 4 p. 245-251.

Clemmons, E.K., 1985, Data models and the ANSI/SPARC architecture, *in* Yao, S.B., ed.,*Principles of database design*: Prentice-Hall, Inc., New York, p.66-114.

Jones, E., 1987, *Using dBase III Plus*: Osborne/McGraw-Hill, New York, 516p.

Kahn, B.K., 1985. Requirement specification techniques, *in* Yao, S.B., ed., *Principles of database design*: Prentice-Hall, Inc, New York, p. 1-65.

CHAPTER 2

The PC-File+ Database Management System: Its Use and Some Examples of Petrologic Applications

Russell G. Clark, Jr.
Dept. of Geology
Albion College
Albion, Michigan USA

2.1 INTRODUCTION TO PC-FILE+

2.1.1 Features and Requirements

PC-File+ is one of the most widely used and inexpensive database management systems available. Although it does have relational capabilities, PC-File+ will be used in this chapter to demonstrate the use of a flat-file type of DBMS. A flat-file, fixed length type of system is analogous to a collection of preprinted forms each of which contains blank spaces which are to be filled in with the appropriate information. After some or all of the blank spaces on a form are filled in, that form is then filed with other forms as part of the database. A relational DBMS allows data to be retrieved from more than one database at a time, whereas a flat-file

DBMS allows the use of only one database at a time. The major advantages of PC-File+ are that it is inexpensive (the current retail price in US dollars is $69.95) and it is easy to use (it is mainly menu-driven and there are 200 context-sensitive help screens).

PC-File+ will run on any of the microcomputers in the IBM series and on virtually all microcomputers which claim IBM compatibility. The program requires a microcomputer with an MS-DOS or PC-DOS (version 2.0 or later) operating system, a minimum of 384K RAM, two double-sided disk drives (or one double-sided drive and a hard disk), and an 80-column display monitor.

There are other versions of PC-File available. These include PC-File, PC-File III, and PC-File/R, but the newer PC-File+ represents a major step forward in the development of good, but inexpensive, database management software. Another recent version is PC-File:db, which is similar to PC-File+, but has the additional ability to read and write dBASE database files.

2.1.2 A Note on Notations

In the following descriptions, text responses which you are to type into the microcomputer will be shown in boldfaced and italicized type, such as ***pcf***. PC-File+ prompts, commands, and functions will be shown in boldfaced type, but will not be italicized (for instance, **Please reply**). When the text refers to an individual keyboard key, the notation for the key will bracketed and boldfaced, so for instance the "F" key would be shown as **<F>**.

When you are required to type a text response into the computer, the typed text should be followed by pressing **<Enter>**. On some keyboards the **<Enter>** key is labeled **<Return>** or may indicated by the symbol ____. In this chapter, the instructions may be explicit ("press **<Enter>**") or may be implied. In any event, if nothing happens after the text has been typed, go ahead and press the **<Enter>** key.

In a few examples in this chapter quotation marks (" ") may be used to delimit some text which is to be typed into the microcomputer, but unless there are explicit instructions to the contrary these quotation marks are not to be typed. Elsewhere in this chapter quotation marks may be used to delimit text which has been displayed by the microcomputer on the screen, but the quotation marks themselves will not appear on the screen.

During the operation of PC-File+ you may be asked to press the **<F10>** function key to indicate that information which has been typed on the screen is correct and should be accepted by PC-File+. In some, but not all, of these situations pressing the **<Enter>** key will have the same effect.

2.1.3 Moving the Cursor and Editing

During the operation of PC-File+ user responses (through typing at the keyboard) are required frequently. The typed response will be displayed on the screen at the current location of the cursor (which is easily recognized as the blinking block). In many situations the user can control where a response will be typed on the screen by moving the cursor to the desired location. The special keys which control the movement of the cursor are described in Table 2.1. Not all of these keys are usable in every situation requiring a user response, but no damage will occur if an inappropriate key is pressed.

PC-File+ normally is in "overtype mode", so that characters can be changed by typing over the old text. The **<Ins>** key acts as a toggle to "insert mode", so that characters are inserted at the cursor position. The **<Del>** and **Backspace** keys perform their normal screen editing functions. Other useful editing functions are described in Table 2.2.

2.1.4 EXTRACT: The Database Example

In describing the use of PC-File+ and its functions it will be helpful to start with an existing database. The EXTRACT database file, which will be used in the exercises later in this chapter, is a database which contains geochemical information about some igneous rocks. The data for this database are included in a comma-delimited file (named EXTRACT.CDF) on the diskette which accompanies this workbook. These data have been taken from a larger igneous rock database named IGBADAT. The file EXTRACT.CDF contains information about 408

Table 2.1 Cursor control keys used with PC-File+

Key	*Explanation*
Arrow keys	Move cursor one space in direction of arrow.
<Home>	Moves cursor to left end of current field.
<End>	Moves cursor to right end of whatever has been typed in current field.
<PgUp>	Moves cursor to first input field on screen.
<PgDn>	Moves cursor to last input field on screen.
<Tab>	Moves cursor to next field on screen.
<Shift><Tab>	Moves cursor to previous field on screen.
<Enter>	Moves cursor to next field on screen, except that if cursor is on last field, then pressing <Enter> is same as pressing <F10> (see next).

samples of granite, andesite and basalt. The PC-File+ database into which you will import the data from the comma-delimited file already has been defined and created and it is located on the same diskette as the comma-delimited data file. After the data have been imported, each of the 408 records in this database will consist of twelve fields. The field names and field lengths for the EXTRACT database are shown in Table 2.3.

2.1.5 File Nomenclature in PC-File+

PC-File+ uses and creates a number of different files. The names of each of these files may consist of up to eight characters followed by a period and then an "extension" which may consist of up to three characters. The three character extension may be used to identify the different types of files. Some of the more common extensions are shown in Table 2.4. Other extensions used by PC-File+ are described in the user manual.

Table 2.2 Special keys used with PC-File+

Key	*Explanation*
<Esc>	Cancels current operation.
<F10>	Used to indicate that all of data on screen OK and to be accepted.
<Ctrl><A>	Same as <F10>.
<Del>	Deletes character at cursor.
Backspace	Deletes character to left of cursor.
<Ins>	Toggles between "overtype mode" and "insert mode" .
<Ctrl><B>	Erases characters from cursor location to right end of current field.
<Ctrl><D>	Duplicates information which was typed for all fields in previously viewed record.
<Ctrl><F>	Duplicates information which was typed for this field in previously viewed record.
<Ctrl><R>	Reads data typed in current field and saves it in memory buffer.
<Ctrl><W>	Writes data saved in memory buffer into current field.
<Alt><H>	Will provide pop-up help window at almost any time during running of PC-File+. This help function is "context sensitive" and therefore useful if one is in doubt as to what to do next.

2.2 LOADING AND STARTING PC-FILE+

For the purposes of the exercises in this chapter, it will be assumed that the user is familiar with DOS and the elementary commands needed to format diskettes and copy files. If not please refer to the user and reference guide which came with your operating system. It also is assumed that the user has installed PC-File+ on a floppy diskette or in a hard disk

Table 2.3 Field names and lengths for the EXTRACT database

Name	*Length*	*Name*	*Length*
GROUP	4	MNO	6
SPEC_ID	2	MGO	6
REF_NO	5	CAO	6
LAT	9	NA2O	6
LON	9	K2O	6
ROCK_NAME	25	P2O5	6
SIO2	6	CO2	6
TIO2	6	H2O_PLUS	6
AL2O3	6	H2O_MINUS	6
FE2O3	6	STATUS_SYM	40
FEO	6		

directory. After the installation process the PC-File+ diskette or hard disk directory should include three files:

(1) PCF.EXE (the executable program file).

(2) PCF.PRO (the main profile file).

(3) PCF.HLP (the context-sensitive help files).

In order to complete the exercises described in this chapter it will be necessary to use some of the files which are on the diskette which accompanies this workbook. The following files should be copied from the workbook diskette to an empty formatted data diskette or to an appropriate data directory on a hard disk:

EXTRACT.DTA
EXTRACT.HDR
EXTRACT.INX
EXTRACT.CDF

Table 2.4 Some common PC-File+ file extensions

Extension	*Explanation*
.exe	"Executable" file. File named PCF.EXE is main PC-File+ program file. It is file which is run when you type ***pcf*** to start PC-File+ program.
.hlp	"Help" file. File named PCF.HLP contains 200 context-sensitive help screens which are accessed when **<Alt><H>** is pressed. If it is to be used, PCF.HLP must exist in same directory as PCF.EXE file.
.dta	"Data" file. Contains actual data which has been entered into database.
.hdr	"Header" file. Contains database definitions such as those defining field names and field lengths which are used in database.
.inx	"Index" file. Keeps track of where each data record is located in database "data" file. Also determines order in which records will be displayed.
.pro	"profile" file. Optional PC-File+ file in which some user-determined default characteristics of a database can be recorded. Includes things like screen colors, default directory paths, printer defaults, and passwords.
.rep	"report" file. After you create output report format for displaying data from PC-File+ database, you are given option of saving the report format for later use. Format will be saved in one of these "report" files.
.ans	"answer" file. Stores most recent user-supplied answers to questions about report formats, so that these answers become default responses.
.cdf	"comma-delimited" file. File EXTRACT.CDF is in "comma-delimited" format and can be imported by PC-File+ in order to create database.

The first three files listed previously are the files which define a PC-File+ database termed EXTRACT. The last file in the list is the comma-delimited file which contains the data which will be imported into the PC-File+ database.

The examples and figures shown in this chapter were created by using Version 2.0 of PC-File+ which had been installed on a hard disk and by using the EXTRACT files (see previous) on a 360K floppy data diskette in drive A.

After you have created your EXTRACT data diskette (or directory), start PC-File+ program by doing one of the following:

(1) If you are loading and running PC-File+ from a floppy diskette, place the diskette in drive A, make drive A the default drive, and type ***pcf***; or

(2) If you are loading and running PC-File+ from a hard disk, make the directory containing the PC-File+ program files (PCF.EXE, etc.) the default directory and type ***pcf***.

When PC-File+ loads and starts to run, its logo and copyright information appear on the screen along with the prompt **"Which drive for Database (A-Z)?"**, as shown in Figure 2.1. You should reply with the letter of the drive which contains the database files. The proper response usually will be to press either **<A>** or **<B>**, if the database file is on a floppy diskette, or to press **<C>**, if the database files are on a hard disk. After the proper drive letter has been entered, a new prompt appears:

"What path for the data? _ _"

If the database files are in the root directory (or if you are not using directories or sub-directories), then leave the response field blank and just press **<Enter>**. If the database files are in any directory other than the root directory, then respond by typing in the full path name for the directory.

The next screen that appears (see Fig. 2.2) lists the names of each of the databases which exist on the drive (and directory) just named. In addition, there is a blank field into which the name of a new database could be

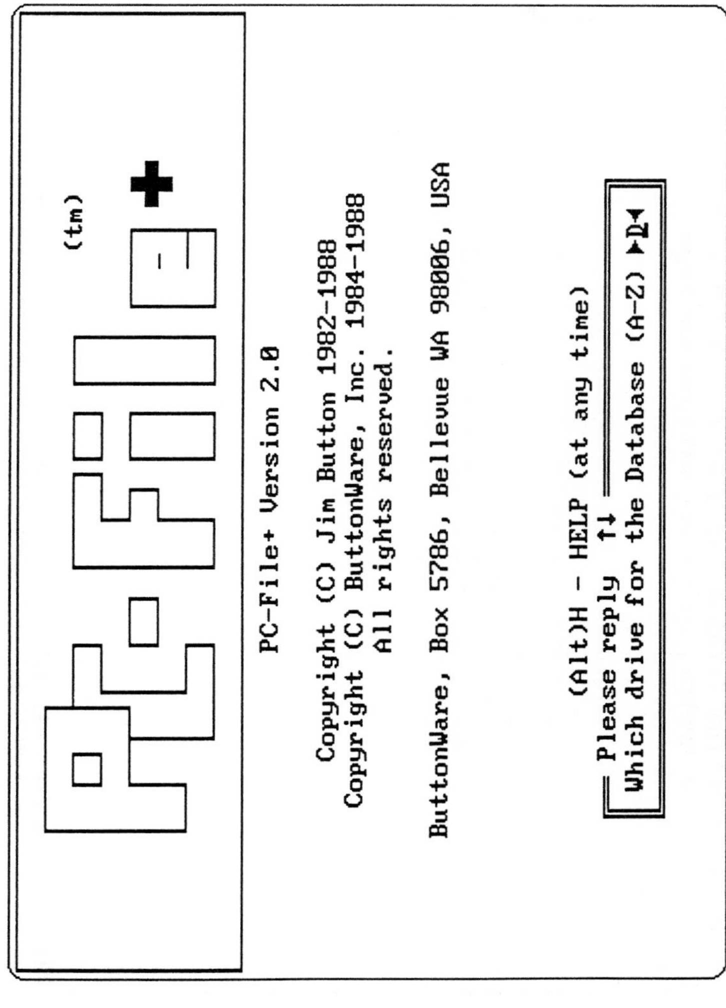

Figure 2.1 PC-File+ title screen and first prompt.

```
Select an existing database or give a name for a new database.
Type a name or number, or select with cursor keys
 ►_        ◄
 [1]  EXTRACT




Please respond.  (F10) when complete.  (Alt)H for help.
```

Figure 2.2 Database Selection Screen.

entered. Press the **<down-arrow>** once to move the cursor to the database named **EXTRACT** (the database example to be used with this chapter) and press **<F10>** to complete the selection.

2.2.1 Importing the EXTRACT data

The next screen (Fig. 2.3) to appear is the initial Master Menu Screen. The name of the database is shown on the second line. The third line ("**There are 0 records**") indicates that the EXTRACT database, although defined and created, is empty. Before the EXTRACT database can be used for the exercises in this chapter, the data contained the comma-delimited file named EXTRACT.CDF must be imported into the PC-File+ database. Because the database currently contains no records, the initial Master Menu Screen shows only a limited number of available functions. After the data in the comma-delimited file has been imported, the Master Menu Screen will show more available functions and it will be described fully later in this chapter.

Press **<F8>** or **<U>** to go to the PC-File+ Utilities Menu. The screen will look like Figure 2.4. Press **<I>** to indicate you want to import a file. The next prompt is:

"Import a PC-File database or other file? (F/O) _O_"

Respond by pressing **<O>** since you want to import a comma-delimited file and not a file which already is in PC-File+ format. This is followed by the prompt:

"Which drive contains the file to be imported? (A-Z) _A_"

Because the data diskette is in drive A, press **<A>**. The program then requests the path specification:

"Which path? _   _"

In this example the proper response simply is to press **<Enter>**, because the data diskette in drive A does not contain any directories. The next screen (Fig. 2.5) allows you to which file (from those on the drive and

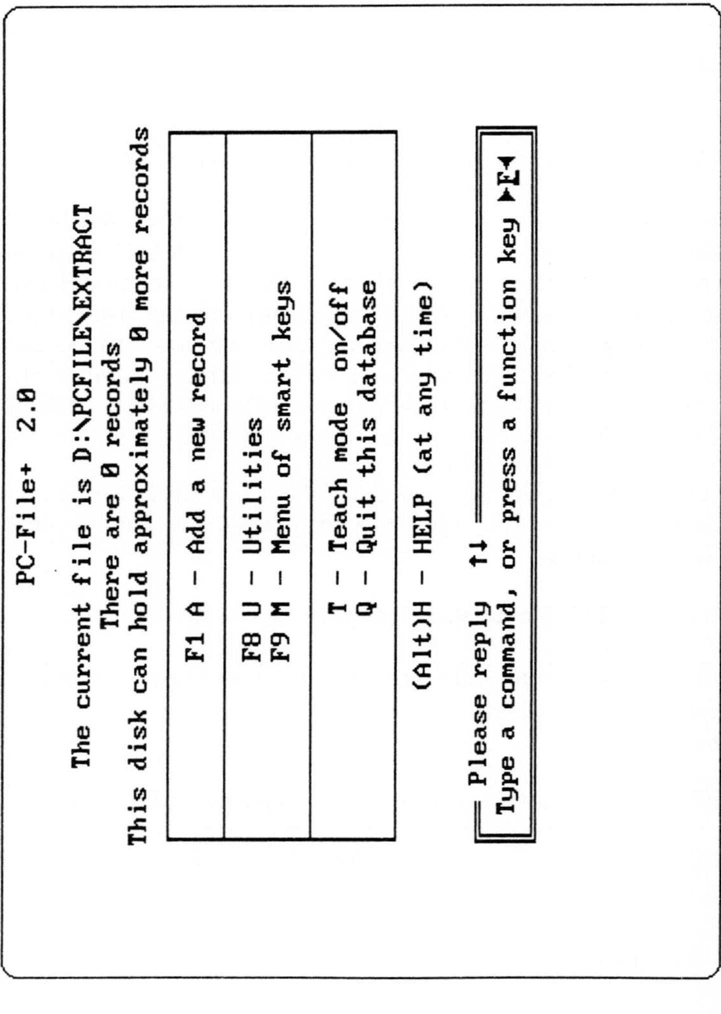

Figure 2.3 Initial PC-File+ Master Menu Screen prior to importing records from comma-delimited data file.

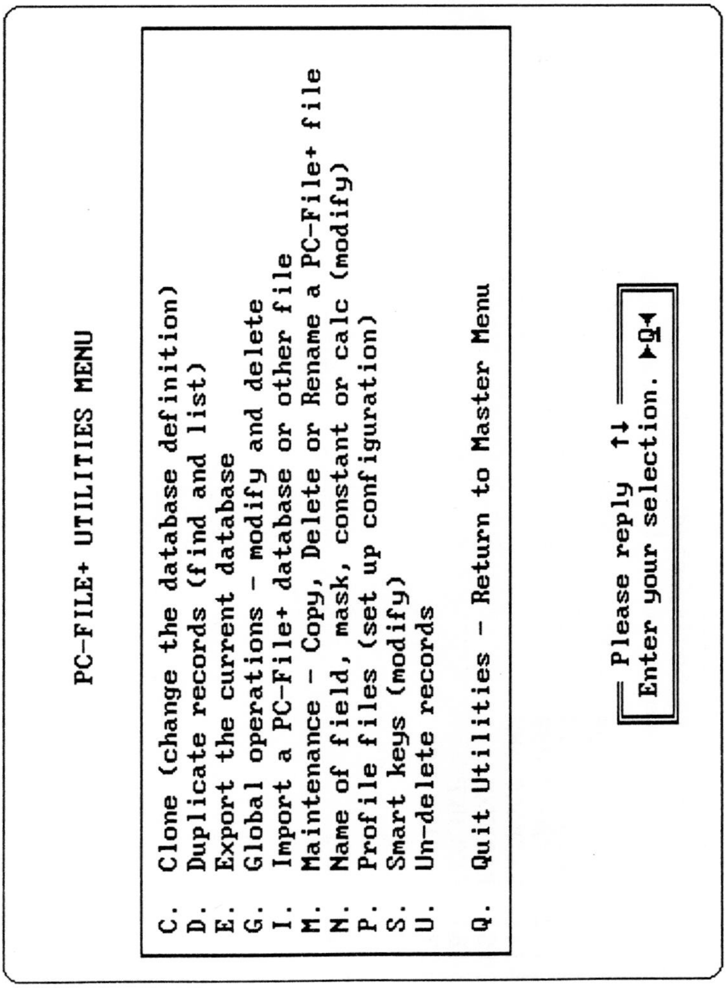

```
PC-FILE+ UTILITIES MENU

C. Clone (change the database definition)
D. Duplicate records (find and list)
E. Export the current database
G. Global operations - modify and delete
I. Import a PC-File+ database or other file
M. Maintenance - Copy, Delete or Rename a PC-File+ file
N. Name of field, mask, constant or calc (modify)
P. Profile files (set up configuration)
S. Smart keys (modify)
U. Un-delete records

Q. Quit Utilities - Return to Master Menu

Please reply ↑↓
Enter your selection. ►Q◄
```

Figure 2.4 PC-File+ Utilities Menu screen.

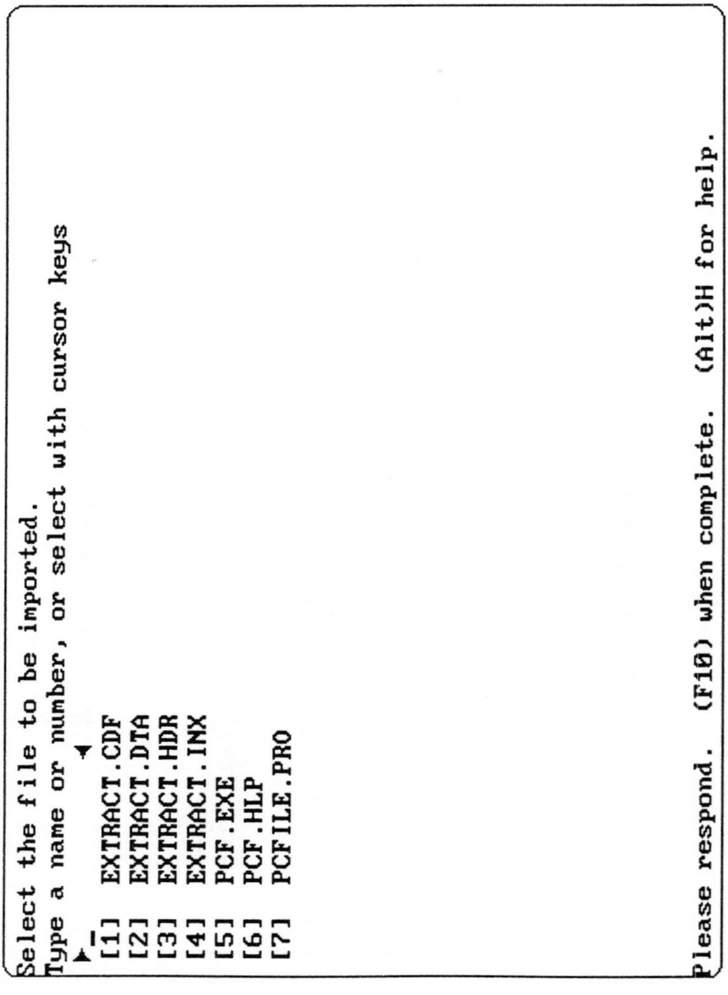

```
Select the file to be imported.
Type a name or number, or select with cursor keys
▶_         ◀
[1]   EXTRACT.CDF
[2]   EXTRACT.DTA
[3]   EXTRACT.HDR
[4]   EXTRACT.INX
[5]   PCF.EXE
[6]   PCF.HLP
[7]   PCFILE.PRO




Please respond.  (F10) when complete.  (Alt)H for help.
```

Figure 2.5 Screen for selecting which file to import.

path indicated above) is to be imported. Move the cursor to the file named **EXTRACT.CDF** and then press **<F10>** to select it. The next screen (Fig. 2.6) requests that you indicate the format of the file to be imported into the PC-File+ database. Press **<M>**, because the file is in comma-delimited format. The next prompt is:

"Import ALL records or SELECTED records? (A/S) _A_"

Whereas it is possible to define criteria to select only part of the EXTRACT.CDF file to be imported into the PC-File+ database, at this time press **<A>** to import all 408 records.

Figure 2.7 shows the results of importing the first record from EXTRACT.CDF into the PC-File+ database. The screen shows each of the field names in capital letters. Note that some of the field names end with a "#" (a "pound-sign"). PC-File+ uses the pound-sign suffix for a field name when that field will contain numeric values. The field lengths and the imported contents for each field are shown between square brackets. Note the prompt at the bottom of the screen. If the field contents of the first record are clearly incorrect (for instance, impossible values for LON, LAT, or any of the oxides), then you probably will want to press **<Q>** to abort the import process, because the format of the imported file is not compatible with the definition of the PC-File+ EXTRACT database. If the contents of the fields appear to be correct and you want to add this record to the PC-File+ database, then press **<Y>**; or you could exclude this record from the PC-File+ database by responding with a **<N>**. In either of these situations, another screen will appear with the contents of the second record, along with the prompt indicating the same four possible responses.

Assuming that everything looks correct, you should now press **<X>** which will cause the import process to proceed without further user intervention. The individual records are no longer displayed as they are imported, but another screen is displayed on which the number of records imported is updated continually. After all 408 records are imported the screen will look similar to Figure 2.8. Press **<Enter>** to complete the import process and return to the PC-File+ Utilities Menu (Fig. 2.4). Then press **<Q>** to return to the PC-File+ Master Menu Screen (Fig. 2.9).

Notice the differences between the Master Menu Screen shown in Figure 2.9 and the one shown in Figure 2.3. The screen now shows that the EXTRACT database is no longer empty ("**There are 408 records**") and as a result several new PC-File+ functions are available for use.

```
                    IMPORT A FILE

  B.  dBASE .DBF files
  C.  PC-Calc                     (by ButtonWare)
  D.  DIF                         (i.e. Visicalc, etc.)
  F.  Fixed length                (i.e. random files)
  M.  Mail-merge                  (comma-delimited)
  P.  Peachtext                   (i.e. 1 field per line)
  T.  Text editor (SDF)           (i.e. fixed length, c/r)
  U.  User-defined field delimiter
  W.  Word Perfect
  X.  Compressed                  (by PC-File export)

   Please reply  ↑↓
  What is the format of the input file? ▶X◀
```

Figure 2.6 Screen for indicating format of file to be imported into PC-File+ database.

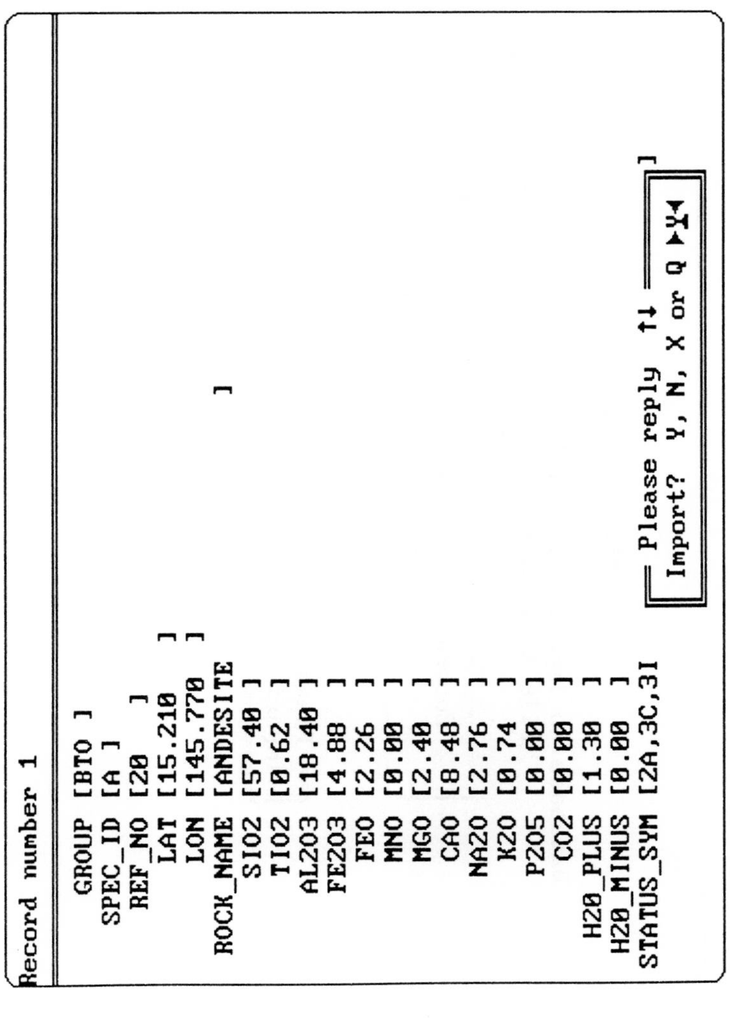

Figure 2.7 First imported record.

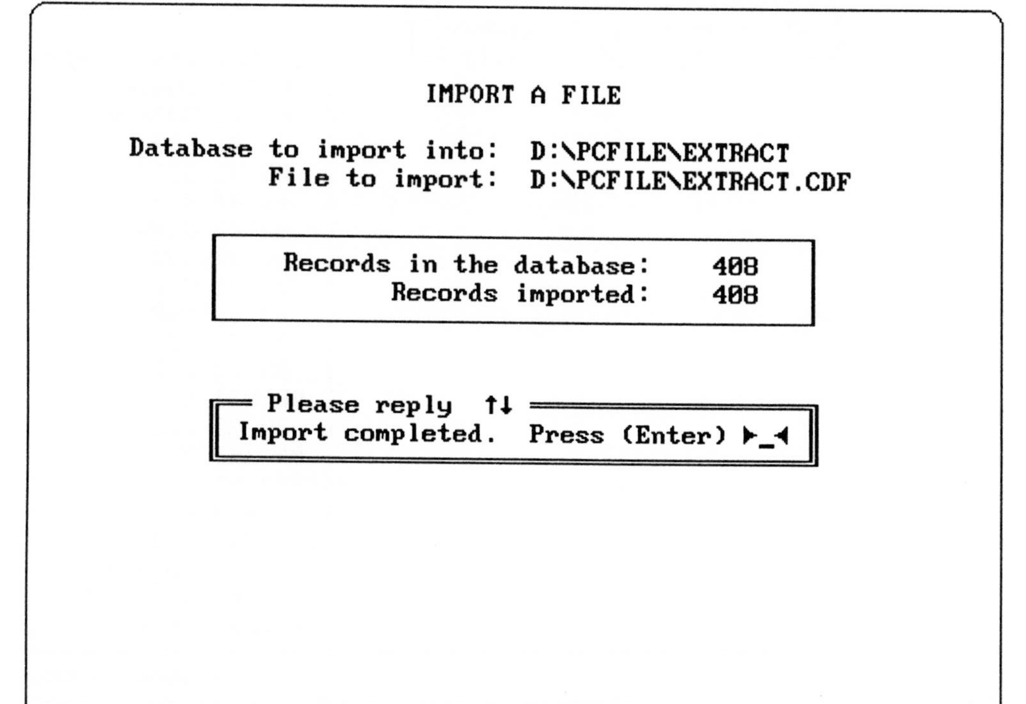

Figure 2.8 Screen indicating that all 408 records from EXTRACT.CDF have been successfully imported into PC-File+ database named EXTRACT.

```
                    PC-File+  2.0

         The current file is D:\PCFILE\EXTRACT
                  There are 408 records
      This disk can hold approximately 0 more records

             F1 A - Add a new record
             F2 F - Find a record

             F4 G - Graphs
             F5 L - Letter writing
             F6 R - Reports
             F7 S - Sort
             F8 U - Utilities
             F9 M - Menu of smart keys

                T - Teach mode  on/off
                Q - Quit this database

            (Alt)H - HELP (at any time)

   Please reply  ↑↓
 Type a command, or press a function key ▶E◀
```

Figure 2.9 PC-File+ Master Menu Screen after importing 408 records from EXTRACT.CDF comma-delimited file.

2.2.2 The Master Menu Screen

The Master Menu Screen (Fig. 2.9) indicates the types of functions which can be performed by PC-File+. This chapter will not attempt to describe all of the available functions, but only those which are considered to be essential or particularly useful for working with the EXTRACT sample database. Those functions which are not described here can be learned either by using the context-sensitive help windows or by reference to the PC-File+ user manual.

The second line of the Master Menu Screen shows which disk drive contains the current database as well as its name. The third line shows the number of records currently in the database. After this database has been sorted at least once, the third line also will indicate the order in which the records have been sorted most recently. The fourth line indicates the calculated number of records which can be added to the current disk. The maximum number of records which can be stored in a particular database depends upon the number of fields and the sizes of those fields. The number and the size of the fields are determined during the definition of a new database. The maximum number of records also depends upon the type of disk being used to store the database; a double-density 360K diskette can store fewer records than a high-density 1.2 Mbyte diskette, which in turn stores fewer records than a hard disk could hold. PC-File+ does not allow a database to continue from one floppy diskette to another, so larger databases (such as the IGBADAT database from which the EXTRACT records were derived) would have to be stored on a hard disk drive.

The outlined portion of the Master Menu Screen contains a table of the available PC-File+ functions. Each function can be initiated by pressing the function key (or letter key) indicated to the left of the function shown on the menu. For example, in order to locate a particular record (or a set of records) the user would begin the process by pressing either **<F2>** or **<F>**. Note that because this is such a generally used function, it is presented as the default function each time the Master Menu Screen appears.

Near the bottom of the Master Menu Screen a line reminds the user that PC-File+ will supply a context-sensitive help screen at any time if the <Alt> and <H> keys are depressed simultaneously. These help screens are especially useful for the beginning user who may be in doubt as to what to do next during a PC-File+ session. Note also that the PC-File+ "teaching mode" can be toggled on and off by pressing the **<T>** key. This mode provides an automatic on-screen help window each time a user response is required.

The most essential and useful PC-File+ functions will be described next. These functions include those which allow the user to add, modify, delete, display, or locate one or more records in a particular database. Other functions which enable the user to change the order in which the records are listed or printed also are essential functions.

2.3 ADDING A RECORD

Databases, such as the EXTRACT database, may be built by adding new records. As indicated in the Master Menu Screen (Fig. 2.9), pressing **<F1>** or **<A>** will begin the process of adding a new record to the database.

A new screen (the Add Screen) is displayed and it will similar to Figure 2.7, except that all of the fields will be blank. At the upper left of the screen the phrase "Record number 409" will appear. When PC-File+ adds a record it searches first for unused space within the sequential records of the database; such unused space would be created where records have been deleted. In the current example none of the 408 records have been deleted from EXTRACT, so the first available space for adding is at record number 409. At the upper right corner of the screen an "A" appears to remind you that PC-File+ is in "add mode". At the bottom of the screen the prompt that appears on the Add Screen will be different than the one shown in Figure 2.7. The Add Screen prompt will read:

"Please respond. (F10) when complete. (Alt)H for help."

At this point you should move the cursor on the Add Screen to the ROCK_NAME field and type in ***DIORITE***. You then can proceed to fill in the information requested for each of the other fields, controlling the cursor location as described in Table 2.1 and keeping in mind that each field can be edited and reedited until the input data are satisfactory. For this record it does not matter what you type into these other fields, because you are going to delete this record later in the chapter anyway. As always, pressing the **<Esc>** key will back up everything by one step; in this example it would return you to the Master Menu Screen.

After the data have been entered satisfactorily into the fields on the Add Screen, there are two ways to accept the data for the new record:

(1) By pressing the **<Enter>** key while the cursor is in the last field; or

(2) By pressing **<F10>** regardless of where the cursor is located.

In either situation the PC-File+ program will display at the bottom of the screen:

"O.K. to ADD? Y,N, or X (to stop asking) _Y_"

This gives the user a chance to check over the input data once more. A response of **<N>** will cause the cursor to return to the input fields and additional changes to the data can be made. A response of **<Y>** (which is the default response) will cause the program to accept the input and present a new blank Add Screen on which the user can type the data for the next record. A response of **<X>** will cause the program to accept the input for this record, but in addition the **<X>** response will cause the program to stop asking the question "**O.K. to ADD ... etc.**" for future records; instead, future records will be added to the database as soon as **<F10>** is pressed.

You should press **<Y>** to accept record No. 409.

After you have completed adding records to a database and a new blank Add Screen has been displayed and nothing has been added to any of the fields, pressing **<F10>** will cause the program to return to the Master Menu Screen. Or, if anything has been added to the Add Screen, you can get out of "add mode" and back to the Master Menu Screen by pressing **<Esc>**.

Do not type anything on the Add Screen for record number 410. Press **<F10>** to return to the Master Menu Screen.

2.4 LOCATING A RECORD

The Find function is used to locate one or more records in the database which match a set of user specified criteria. These criteria are defined on the basis of the contents of one or more fields within the database.

Starting with the Master Menu Screen (Fig. 2.9), pressing **<F2>** or **<F>** (or **<Enter>**, because **<F>** is the default response) will cause the Find Menu Screen to appear (as shown in Fig. 2.10). This menu is used to indicate the general type of search to be used. The type of search is selected by pressing the appropriate key. Pressing any of the keys except **<S>** will cause an immediate display of one or more records from the database. Notice that there are options for going immediately to the first and last records in the database, or to the next or previous record from the one displayed most recently. Pressing **<R>** allows one to go to a specific record if the record number (relative sequential position in the database) is known. Selection of either of the two browse modes will result in a tabular display of the first several fields for 20 records at a time.

The most versatile search method, however, is selected by pressing the <S> key. Note that this is the default response for this screen. Press **<S>**, you then will be asked whether you want to do a "Simple Search" or a "Complex Search".

"S(Simple search), C(Complex search), Q(Quit) S/C/Q _S_"

2.4.1 Simple Search

The screen for defining the Simple Search criteria (Fig. 2.11) lists the field names and shows the field lengths as the previous screens have, but in this screen the fields are filled in entirely with asterisks (the symbol "*"). The asterisks are used instead of blank spaces so that spaces can be included in the search criteria. A Simple Search criterion is defined by typing the data to be matched into one of the fields.

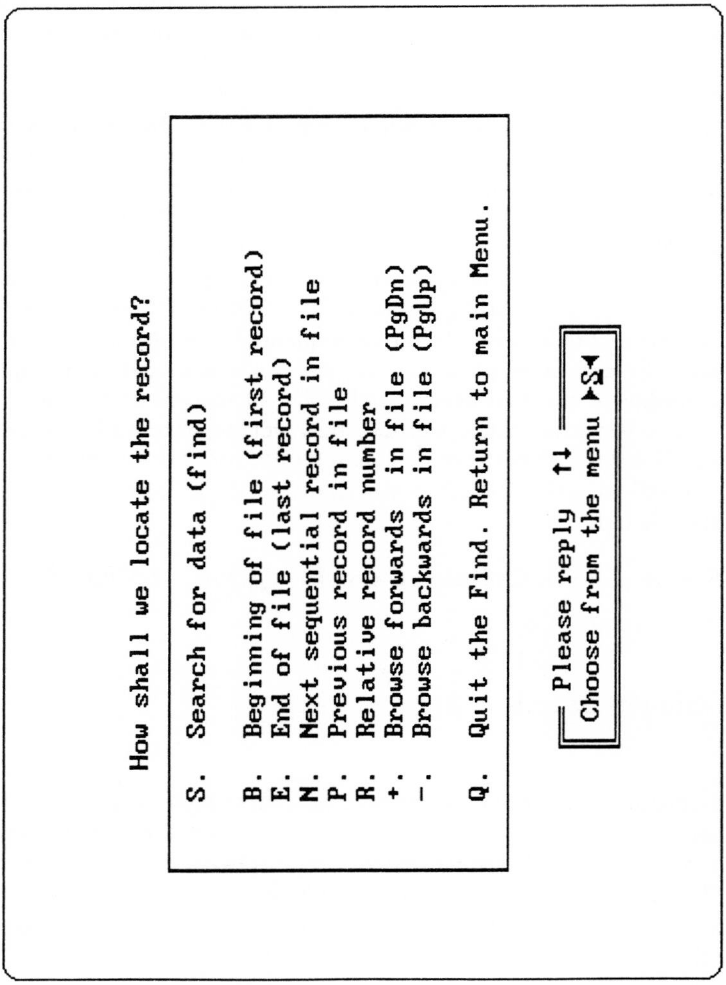

```
How shall we locate the record?

S.  Search for data (find)

B.  Beginning of file (first record)
E.  End of file (last record)
N.  Next sequential record in file
P.  Previous record in file
R.  Relative record number
+.  Browse forwards  in file (PgDn)
-.  Browse backwards in file (PgUp)

Q.  Quit the Find. Return to main Menu.

Please reply ↑↓
Choose from the menu ►S◄
```

Figure 2.10 Find Menu Screen.

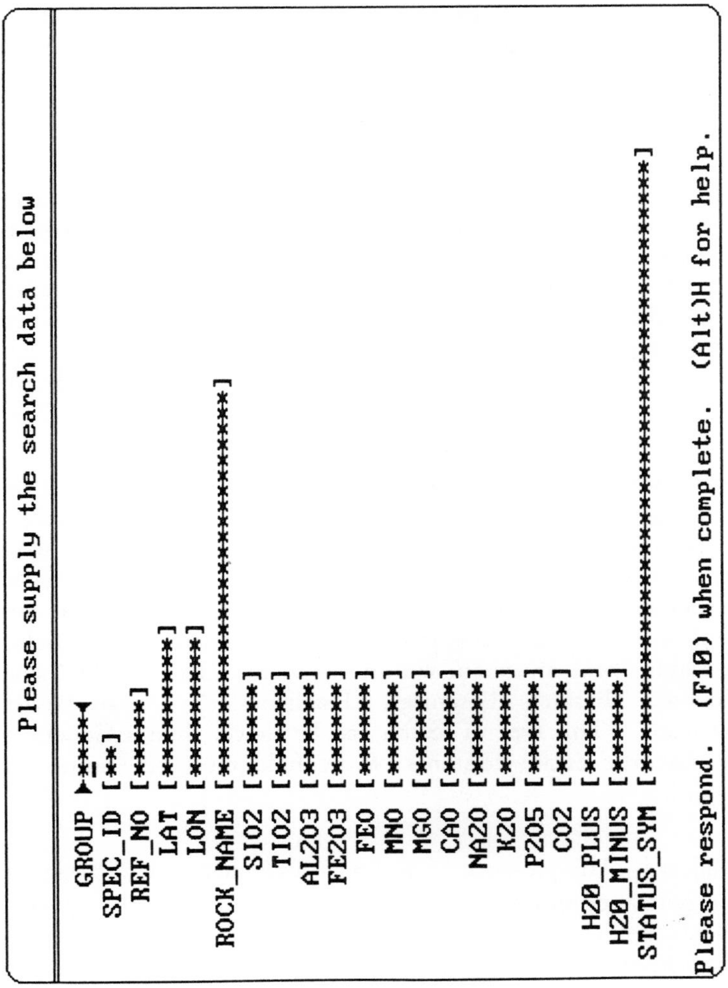

```
                 Please supply the search data below

      GROUP ▶*****◀
    SPEC_ID [**]
     REF_NO [*****]
        LAT [**********]
        LON [**********]
  ROCK_NAME [*************************]
       SIO2 [*******]
       TIO2 [*******]
      AL2O3 [*******]
      FE2O3 [*******]
        FEO [*******]
        MNO [*******]
        MGO [*******]
        CAO [*******]
       NA2O [*******]
        K2O [*******]
       P2O5 [*******]
        CO2 [*******]
  H2O_PLUS  [*******]
  H2O_MINUS [*******]
 STATUS_SYM [****************************************]

Please respond.  (F10) when complete.  (Alt)H for help.
```

Figure 2.11 Simple Search Screen.

There are four different ways in which Simple Search criteria can be entered into the fields:

(1) **Generic.** Whatever is typed in the field must be matched exactly before a record will be selected from the database. If the search criterion is four characters in length, then the selection from the database will be based on the first four letters of each entry in that particular field. For example, if ***Gran*** is typed in a particular field, a search through the database based only on that one field will select all records containing an entry in that field which begins with "Gran". This would include, for example, "Granite", "Granodiorite", and "Granitoid", but not "Leucogranite" or "Biotite Granite". Note that the default condition in PC-File+ is that database searches are not case-sensitive (although it is possible to change the default condition and make them case-sensitive) so that in this example fields beginning with "gran", "GRAN", or even "gRAn" would all be selected. This type of search results in the fastest selection of records from the database.

(2) **Scan Across.** This type of search occurs when the data typed in a field begins with a "~" (a tilde). This will cause a particular record to be selected if the indicated characters occur anywhere within the field entry for a record. For example, if ***~Bas*** is typed in a particular field, a search through the database based only on that one field will select all records containing an entry in that field which contains "Bas" anywhere within the field. This would include, for example, "Basalt", "Basanite", "Trachybasalt", "Alkali Basalt", and "Diabase". This type of search can be slower than a Generic Search, because the entire contents of the field must be scanned in each record.

(3) **Wildcard.** The presence of an underscore character ("_") anywhere in the search criterion causes a Wildcard Search. This type of search selects records which have specific characters at certain specified positions with the field. The "_" in the search criterion indicates that any character can occupy that position within the field. If, for example, a particular field consisted of a two-digit number and each digit held some specific coded information (a "1" in the second

position might indicate "igneous", a "2" might mean "metamorphic", etc.), then typing ***_1*** in as the search criterion would cause the program to select all records with a "1" in the second position regardless of what character was in the first position. Note that wildcards (underscore characters) may be used in combination with both the Generic and Scan Across types of searches.

(4) **Sounds-Like.** If the search criterion in a field begins with a question mark ("?"), then a Sounds-Like Search occurs, which means that the database is searched for all records with data in a particular field which "sounds like" the data described in the search criterion. This can help locate records which may have been entered into the database with misspelled words or with words which may be spelled in multiple ways. For example, if ***?granite*** is entered as the search criterion for a particular field, then records which contain "granite", or "granit" (German), or "granito" (Spanish), or the misspelled "grannite" or "granet" in that specified field all will be selected from the database. One needs to be a little cautious using this mode, however, because unexpected records can be selected. For instance, the same ***?granite*** criterion also would result in the selection of records which contained "granodiorite" or even "garnet" in that field.

Although the examples cited each involved a database search based on the criterion typed into only one field, in fact, a Simple Search for selected records can simultaneously be based on multiple fields. Note also that whereas a Simple Search may be useful for searching on fields which contain text, it is not useful for searching on numeric fields. Regardless of the type of Simple Search performed, the search is initiated (after the search criteria have been defined) by pressing **<F10>.**

2.4.2 Complex Search

If a Complex Search is selected, then the next screen will look similar to Figure 2.12. In order to do a Complex Search the user describes the search criteria in a formula which can include comparison operators, logical operators, data identifiers, and parentheses for grouping items.

Comparison operators include:

>	greater than
<	less than
=	equal to
!=	not equal to
>=	greater than or equal to
<=	less than or equal to

Logical operators include:

&	AND
|	OR

Data identifiers can include:

(1) Field names. The full list of field names is listed on the Complex Search Criteria Screen (Figure 2.12). It is not necessary to use the full field name, only enough of the first part of the name to distinguish it from other field names. For example, Si could represent a field named SIO2 as long as there were not any other field names which began with the letters "Si".

(2) Record numbers. Each record number (indicating the record's sequential location in the database) is indicated by a # character. For example, if (# <= "250") were part of the selection criteria, then only the first 250 records would be included in the search.

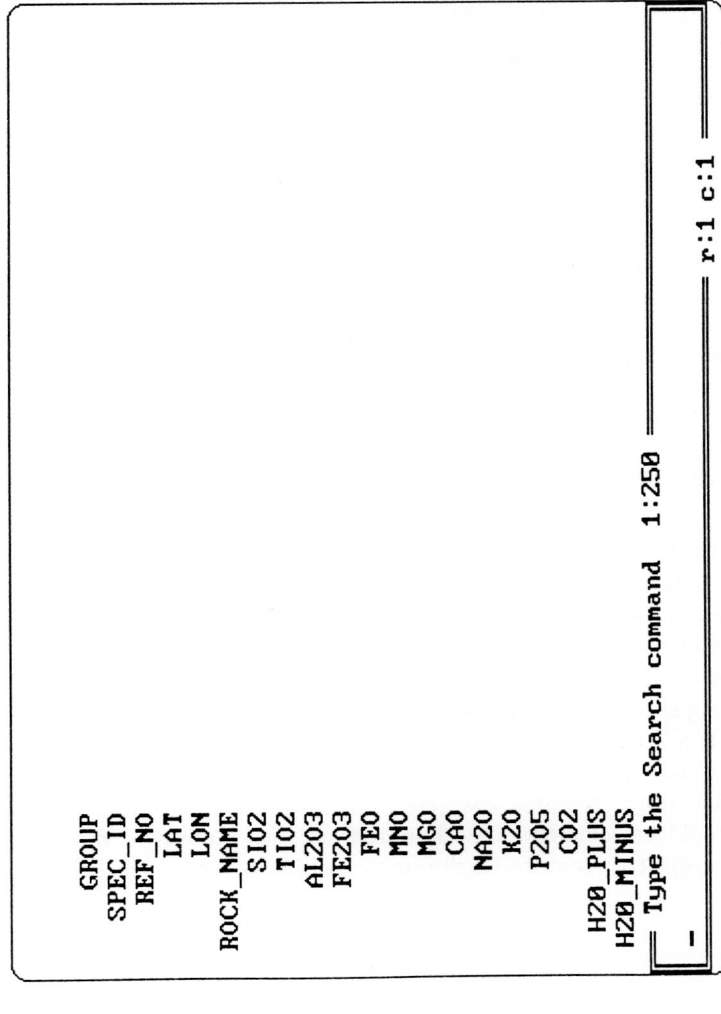

Figure 2.12 Complex Search Screen.

(3) Constants. Some constants are similar to the data used for comparisons and typed into the fields for a Simple Search. Generic constants are surrounded by quotes (e.g. "Gran"). Scan Across constants are surrounded by tildes (for example, ~Bas~). Sounds-Like constants are surrounded by question marks (for example, ?Granite?). Note that wildcard characters (underscore characters) can be used in both the Generic and Scan Across types of constants. In addition, Numerical constants may be compared to the contents of any numerical field (for example, Si>70)

The Complex Search criteria are described in a formula using the appropriate identifiers and operators along with whatever parentheses are necessary to establish the proper logic for comparison and selection of database records. This Complex Search criteria formula may be up to 250 characters long and is typed in the box at the bottom of the Complex Search Criteria Screen (Fig. 2.12).

Note the following examples (using the EXTRACT database) of the Complex Search criteria formulae:

(1) ***(ROCK = "Granite ") & (SI < 72)*** would result in the selection of all records for which the rock name was recorded as granite and the silica content was < 72%. Note that in order to be selected both criteria would have to be met. Forty-two records in the EXTRACT database meet these criteria. The first record determined as a result of this Complex Search is record No. 15 and is shown in Figure 2.13.

(2) ***(ROCK = "Granite ") | (SI > 72)*** would select all records for which either one or the other of the criteria was met. That is, all records for which the rock name was granite or for which the silica content was > 72% would be selected. One hundred and two records in the EXTRACT database meet these criteria, the first of which is record No. 14.

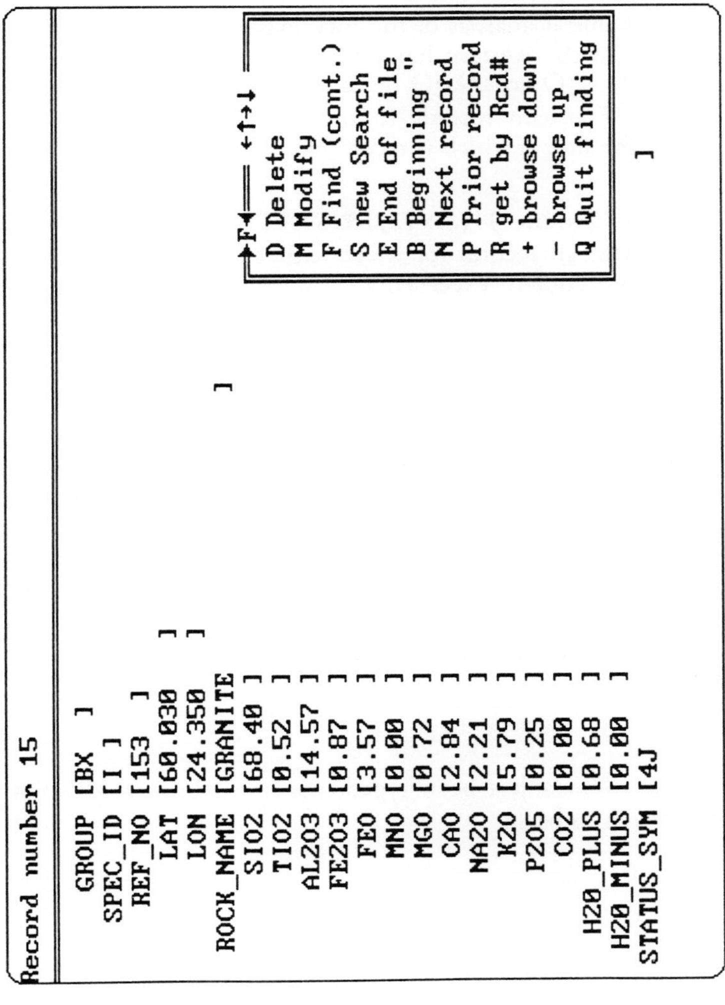

Figure 2.13 Find Screen, which is displayed after search has been completed.

(3) ***(ROCK != "Granite ") & ((K+NA) > 6)) & (LON < 0)*** would select all records for which three criteria were met: the rocks were not named granite; and the sum of the alkali oxides (K_2O + Na_2O) from the geochemical analysis of the rock exceeded 6%; and the rocks were collected in the western hemisphere. Note that for each record the sum of the K_2O and Na_2O fields was calculated as part of the search criteria. Fourteen records in the EXTRACT database meet these criteria, the first of which is record No. 108.

(4) ***((ROCK="Granite") | (SI>72))&(STAT!=~2B~)& (STAT!=~2D~)*** would select all records for which either the rock was named granite *or* the silica content exceeded 72. In addition, this search would only select records whose STATUS_SYM list did not contain the symbol "2B" (i.e. omit records for which the analysis was normalized to 100% in the source reference) and did not contain the symbol "2D" (i.e. omit records for which iron is given as "total iron" in the source reference). Notice that the blank spaces which were used in the first three examples served to make those search formulae more readable, but that the blanks are not necessary. Seventy-one records in the EXTRACT database meet these criteria, the first of which is record No. 14. Comparison of this formula (and these results) to example (2) previous indicates that there are 31 records in the EXTRACT database for which: (a) the rock name is either granite, or the silica content exceeds 72%, and (b) which contain either a "2B" or a "2D" status symbol.

After the formula has been typed, pressing **<Enter>** or **<F10>** will cause the search to begin.

2.4.3 After the Search

After either a Simple or a Complex type of search, the first of the records which met the criteria will be displayed on the Find Screen similar to the screen shown in Figure 2.13. Notice the pop-up command window which appears in the lower right corner of the screen. The "F" at the top of this window reminds you that Find is the currently active function. The window lists most of the PC-File+ commands which are available for use at this point. If necessary, it is possible to move the window to another portion of the screen by using the arrow keys. The next record selected during the search (i.e. the next sequential record in the database which meets the defined selection criteria) may be viewed by pressing **<F>** (or **<F10>** or **<Enter>**). Pressing **<S>** initiates a new search and allows you to define new selection criteria. The keys **<E>, <B>, <N>, <P>, <R>, <+>,** and **<->** each have the same result as previously described in association with Figure 2.10, but it is important to note that the records displayed using these keys are drawn from the total database and not just from the records selected most recently. Therefore, for example, if record No. 16 is currently displayed and you press **<N>**, then the record displayed next will be record No. 17, even though record No. 17 does not meet the selection criteria which were used to display record No. 16 (although record No. 17 is a granite, it does not have a silica content of < 72%). Pressing **<Q>** (or pressing **<Esc>**) returns the user to the Master Menu Screen. Note that it also is possible to modify or delete the record displayed as the result of a search. These two commands will be described later in this chapter.

2.4.4 Browse Mode

The initial Find Screen (Fig. 2.13) displays one record at a time on the screen. Browse mode can be used to view 20 records at a time. Pressing **<+>** from the Find Screen displaying record No. 15 results in a Browse Screen which will look like the one shown in Figure 2.14. As with other

	GROU	SP	REF_N	LAT	LON	ROCK_NAME	SIO2
16	BX	H	153	60.050	24.020	GRANITE	74.76
17	BX	G	153	61.400	23.980	GRANITE	73.03
18	BX	E	153	60.220	23.550	GRANITE	71.69
19	BX	D	153	61.600	22.600	GRANITE	66.18
20	BX	B	153	60.470	21.080	GRANITE	72.63
21	QB	D	154	44.950	-72.530	GRANITE	73.41
22	QP	C	157	41.000	-121.000	BASALT	48.50
23	CC	A	162	0.000	30.000	GRANITE	64.38
24	DJ	C	199	-32.990	151.250	BASALT	46.80
25	DJ	B	199	-32.990	151.250	BASALT	42.07
26	DJ	A	199	-32.990	151.250	BASALT	45.63
27	DL	N	201	46.400	-117.100		
28	DL	M	201	44.100	-116.500		
29	DL	L	201	44.100	-118.200		
30	DL	K	201	44.100	-116.500		
31	DL	J	201	44.100	-118.200		
32	DL	I	201	47.630	-120.070		
33	DL	H	201	44.100	-118.200		
34	DL	G	201	45.600	-117.500		
35	DL	F	201	45.600	-117.500		

```
►Q◄  ←↑→↓
S new Search
E End of file
B Beginning "
N Next record
P Prior record
R get by Rcd#
+ browse down
- browse up
Q Quit finding
```

Use (Tab) and (Shift)(Tab) to position the fields

Figure 2.14 Browse Screen which would be displayed if <+> were pressed while viewing FIND Screen in Figure 2.13.

pop-up windows used in PC-File+, the window on the Browse Screen can be moved to another screen location by using the arrow keys. Note that each database record is displayed on the screen in one row and thus the data for a particular field for all 20 records occur in a single column. Because the total screen width is only 80 characters, the Browse Screen can display approximately 75 characters from each record. However, any remaining fields for these 20 records can be displayed by pressing the **<Tab>** key; each press of the **<Tab>** key moves the "browse window" one field (column) to the right. Pressing simultaneously the **<Shift>** and **<Tab>** keys causes the browse window to move back to the left, moving one field for each key press.

Browse mode can be entered directly from the Find Screen by pressing **<+>** or **<->** or by pressing the **<PgDn>** or **<PgUp>** keys. Once in browse mode pressing <+> or <PgDn> will cause the next 20 records in the database to be displayed, while pressing **<->** or **<PgUp>** will cause the previous 20 records to be displayed. Pressing **<Enter>** or **<F10>** causes the program to return to the Master Menu Screen rather than to the screen from which browse mode was entered. If you want to exit from browse mode and return to the Find Screen press **<F>.**

It is not possible to make changes to the database while it browse mode; it is designed solely for a quick viewing or scanning of the database.

2.5 MODIFYING A RECORD

After a particular record has been added to a database, it may become necessary (resulting from an error or new data, for instance) to change the contents of one or more fields in that record. After a record has been located using the Find function, PC-File+ allows for the modification of any or all of the fields for that record. As indicated in the pop-up command window on the Find Screen (Fig. 2.13), pressing **<M>** will enable you to modify the displayed record and will result in the cursor moving to the first field of the record. The pop-up command window disappears temporarily and "**MOD**" appears at the top right of the screen to remind you of the current function. Then the normal editing functions and keys (Table 2.2) may be used to edit any of the record's fields. A message at the bottom of the screen reminds you that after all modifications have been made, pressing **<F10>** will save the changes, return you to the Find Screen, and bring the pop-up command window back to the screen.

2.6 DELETING A RECORD

After a record has been located using the Find function, the record may be deleted from the database, if necessary. As indicated in the pop-up command window, pressing **<D>** will initiate the deletion of the current record. The pop-up command window disappears temporarily and "**DEL**" appears at the top right of the screen to remind you of the current function. Then a small box appears at the bottom of the screen containing the prompt

"O.K. to delete? Y/N _N_"

The default response is "No", so you will have to press **<Y>**, if the record is to be deleted. If you press **<N>** or accept the "No" default by pressing **<Enter>**, PC-File+ will return you to the Find Screen with the record unchanged, and bring the pop-up command window back to the screen. If you press **<Y>**, the word "**DELETED**" appears to the right of the record number at the top of the screen, and the pop-up command window returns.

The deletion process is not immediately permanent. Deleted records are retained in a PC-File+ database until they are replaced by new records or until the database has been modified permanently by running one of the other utility routines. If you change your mind and want to "un-delete" a deleted record, go back to the Master Menu Screen and press **<U>** which will bring up a menu which lists the available PC-File+ utility routines. The Un-Delete routine will search the database for existing deleted records and then allow the option of restoring each located record to the database.

Now it is time to delete the previously added record No. 409 (the diorite). From the Master Menu Screen press **<F>**. If the database has not been rearranged, you could select the desired record by pressing **<E>** (since it was the last record at the time it was added) or you could press **<R>** which causes the following prompt to appear:

"What record number to find: _ _"

Typing **409** in the space provided also would select the desired record. If the database has been rearranged, then you could locate the record by knowing that it is the only diorite in the database. Press **<S>** on the Find Menu Screen (Fig. 2.10) and then press **<S>** again which will bring up the Simple Search Criteria Menu (Fig. 2.11). Move the cursor to the ROCK_NAME field and type ***diorite***. Then press **<F10>** to initiate the search. When the diorite record appears on the screen, press **<D>** followed by **<Y>** to delete the record.

2.7 SORTING THE DATABASE

When records are displayed on the screen, or sent to a printer or disk file, the sequence in which the records are listed is the same as the sequence in a file created by PC-File+ and termed the "database index". This index file keeps track of where each record is located in the main database file. When a PC-File+ database is created the sequence in which the records are numbered (and displayed) is determined simply by the sequence in which the records were typed (or imported) into the database. In the example of the EXTRACT database this sequence happens to be in order of increasing reference number (the contents of the REF_NO field. The sequence in which records are listed can be controlled by changing the sequence in the database index and this is accomplished by using the Sort function. For example, it might be desirable to have reports listed with the sample identification labels in alphabetical order. For the EXTRACT database this can be achieved by sorting the database index based on the contents of the GROUP and SPEC_ID fields. Note that it is not the database itself which is sorted, but rather the database index, which is why the sort can be accomplished quickly. Note also that it is possible to sort the index on more than one field at a time. In fact, it is possible to sort on as many as ten fields at a time.

To begin the Sort function press **<S>** or **<F7>** while viewing the Master Menu Screen. A list of field names will be displayed on the next screen along with a prompt line at the top of the screen:

"Sort field #1 _   _".

The primary sort field can be selected by typing in the field name (or enough of the name to identify it uniquely) or alternatively by using the arrow keys to move the cursor to the desired field name in the list. In either situation after the field is selected, pressing **<F10>** will cause the field to be accepted. A "1" appears to the left of the field name in the list and another prompt line appears at the top of the screen:

"**Ascending or Descending (A or D) _A_**"

Selecting an "Ascending" sorting order will result in a normal alphabetic order and an "**A**" will appear next to the "1" to the left of the selected field name.

If the sorting will be done on more than one field, the fields should be identified in order of their importance with the most important selected first.

It is also possible to sort on just a portion of a field. This is accomplished by typing (in response to the "**Sort field #**" prompt) the field name, the offset, and the length of the portion of the field on which the sort should be based. For example, in another database there might be a field named SAMPLE_ID with a field length of ten characters. The ninth and tenth positions of the SAMPLE_ID field might contain two digits which represent the year during which the sample was collected. Typing ***SAMPLE_ID,9,2*** in response to the prompt would cause the records to be sorted by year of collection. Sorting based on calculations involving one or more fields is another PC-File+ feature. In response to the "**Sort field #**" prompt, if you typed ***(K+NA)*** instead of a field name, the resulting sort would be based on the sum of the contents of the K_2O and Na_2O fields for each record.

Sorting a database index and rebuilding a new index takes time, but it would take more time to sort the main database file. Larger databases take longer to sort, and the sorting time increases with the number of sort fields which are defined. Sorting time also depends on the sizes of the fields being used. When doing an alphabetic sort, for instance, the time it takes to sort a database index can be shortened by sorting on just the first five or six characters rather than the entire field.

Just before the sorting process begins a prompt at the top of the screen requests that the user name a "work drive" to be used during the sorting. Sorting a database index requires extra storage space, because there must be room for both the original and the newly sorted database index during the sorting operation. During a sort PC-File+ first uses extra RAM memory as work space, and sorting which occurs here takes place quickly. But if the amount of free memory available is small or the database is large, then additional work space is needed and this is when the specified work drive is used for temporary storage. In general, it is not a good idea to put the work drive on the same diskette as the database itself. A better option for a work drive is to use a second diskette in a second disk drive during the sort. If the database is stored on a hard disk, work space is likely not going to be a problem. As the sorting process progresses the status of the sort is reported and updated continually on the screen until the sort has been completed.

2.8 CREATING REPORTS

Reports present the records extracted from a PC-File+ database, or portions of such records, or the results of calculations based on the field contents of such records. These reports can be displayed on the screen, printed on a printer, or saved in a disk file. PC-File+ has a variety of report writing capabilities, which range from the simple to the complex.

Reports are initiated from the Master Screen Menu by pressing **<R>** or **<F6>**. If there are previous report formats which have been saved to disk (files with the ".rep" extension), a list of these saved formats will appear on the screen. To select one of these formats move the cursor to the report format name and press **<F10>**. To define a new format press **<F10>** while the cursor is on the empty response cell. If you are going to define a new report format, or if there are not any saved report formats, then the next screen which is displayed is the Report Format Menu Screen (Fig. 2.15) and it offers a selection of different types of reports, based on the way in which the report will be formatted. Pressing **<P>** will select a "page" format, which will result in each record being displayed on a separate page. Pressing **<R>** will select a "row" format, in which the contents of each record are displayed on a single line with the fields lining up in columns, but because the length of each displayed line is limited (by the capabilities of the printer or display screen), it may be that not all fields can be shown at once. This format is a good one for producing tables of geochemical analyses, although the result is one rock analysis per row rather than the more usual type of tabulation involving a

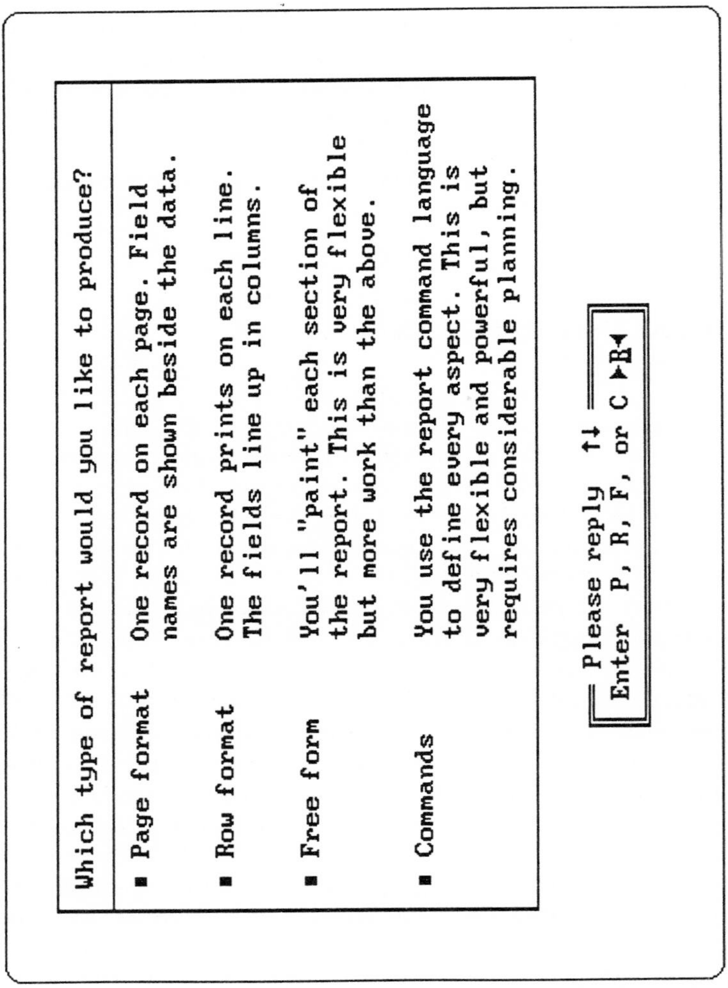

```
Which type of report would you like to produce?

■ Page format    One record on each page. Field
                 names are shown beside the data.

■ Row format     One record prints on each line.
                 The fields line up in columns.

■ Free form      You'll "paint" each section of
                 the report. This is very flexible
                 but more work than the above.

■ Commands       You use the report command language
                 to define every aspect. This is
                 very flexible and powerful, but
                 requires considerable planning.

        Please reply  ↑↓
    Enter  P, R, F, or C ▶R◀
```

Figure 2.15 Report Format Menu Screen.

rock analysis in column format. The "free-form" and "command" types of formats are more versatile than either of the others and allow easier display of computed results (such as averages), but also are much more time-consuming to learn and use. Consequently, these two more complex format types will not been covered here. The interested user is referred to the PC-File+ user manual for further discussion of "free-form" and "command" report formats and use.

2.8.1 The Page Format

Press **<F6>** or **<R>** on the Master Menu Screen. Because there are no report formats which have been saved yet, the next screen is the Report Format Menu Screen. Press **<P>** on the Report Format Menu Screen (Fig. 2.15) to begin a Page Report. You then are asked:

"Would you like to save this report format? Y/N _N_"

Respond by pressing **<Y>**. The next prompt is:

"Enter a name for the format _   _"

As is the situation with any DOS filename, this must be eight characters or less. Type ***PAGFMT1*** in the space provided and press **<Enter>**. The next screen to appear is a Report Menu (Fig. 2.16) which offers various report options. The options listed may be changed by moving the cursor up and down to the desired options and pressing the appropriate key to effect the desired change. For this exercise you should use the options indicated in Figure 2.16. For now the most important options are that the first option be set to "S" (send the report to the screen, rather than to a printer, or disk file) and the last option be set to "S" (print selected, rather than all, records). After you have selected the proper options press **<F10>** to proceed.

Because you indicated that you wanted to have only selected records appear in this report, the next two screens are "Find" screens in which you are able to indicate the type of search (Simple or Complex) and then set the selection criteria just as you did before in the section on "Finding a Record". You should go ahead and set any criteria you like and then

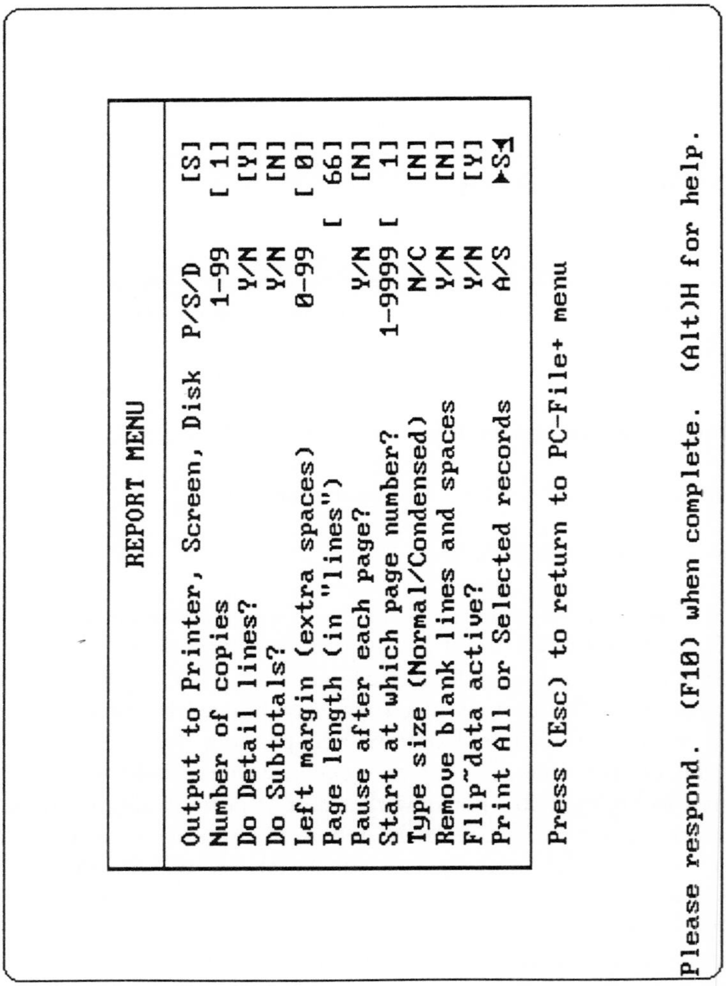

```
                    REPORT MENU

Output to Printer, Screen, Disk  P/S/D      [S]
Number of copies                  1-99    [ 1]
Do Detail lines?                   Y/N      [Y]
Do Subtotals?                      Y/N      [N]
Left margin (extra spaces)        0-99    [ 0]
Page length (in "lines")                [  66]
Pause after each page?             Y/N      [N]
Start at which page number?     1-9999 [   1]
Type size (Normal/Condensed)       N/C      [N]
Remove blank lines and spaces      Y/N      [N]
Flip~data active?                  Y/N      [Y]
Print All or Selected records      A/S      ▸S◂

Press (Esc) to return to PC-File+ menu

Please respond.  (F10) when complete.  (Alt)H for help.
```

Figure 2.16 Report Menu which offers various report options.

press **<F10>** to start the report. Notice that the screen display for the page format only allows a maximum of fifteen fields to be shown at one time. In order to display the remaining six fields for an EXTRACT record, it is necessary to press **<Enter>** and once the remaining fields are displayed, it is not possible to immediately redisplay the first fifteen fields for that record. Pressing **<Enter>** again takes you to the next record selected for the report. Of course, if this report were being sent to a printer there would not be a problem, since each record would be printed on one page. Use of the "Free-form" or "Command" type of report format also could help here, because it allows you to move the position of fields to any location on the screen. See the PC-File+ user manual, or use the help screens, to learn more about "Free-form" and "Command" format reports.

You can view each of the reported records in turn by continuing to press **<Enter>** when indicated, or you can escape from the reporting process by pressing **<Esc>** three times. In either situation you will end up back at the Master Menu Screen. If you continue to press <Enter> until all of the records have been displayed, prior to returning you to the Master Menu Screen, a screen is displayed which indicates the total number of records retrieved for the report as well as a description of the search criteria which were used.

2.8.2 The Row Format

Press **<F6>** or **<R>** on the Master Menu Screen. The next screen asks if you want to define a new format or use a format defined previously for your report. The only report previously defined (PAGFMT1) is shown, but because you will be defining a new format leave the cursor in the empty field and press **<F10>**. Then press **<R>** when the Report Menu Screen seems to indicate that you want to use the Row Format for your report.

The next screen (Fig. 2.17) allows you to select which fields will be included in the report and also the order in which the fields will be printed. Use the cursor to move to each field to be included and number the fields as they will be shown. In the example to be used here, note that the left most column in the report will contain the GROUP data, then the SPEC_ID data, followed by the rock name and five selected oxides.

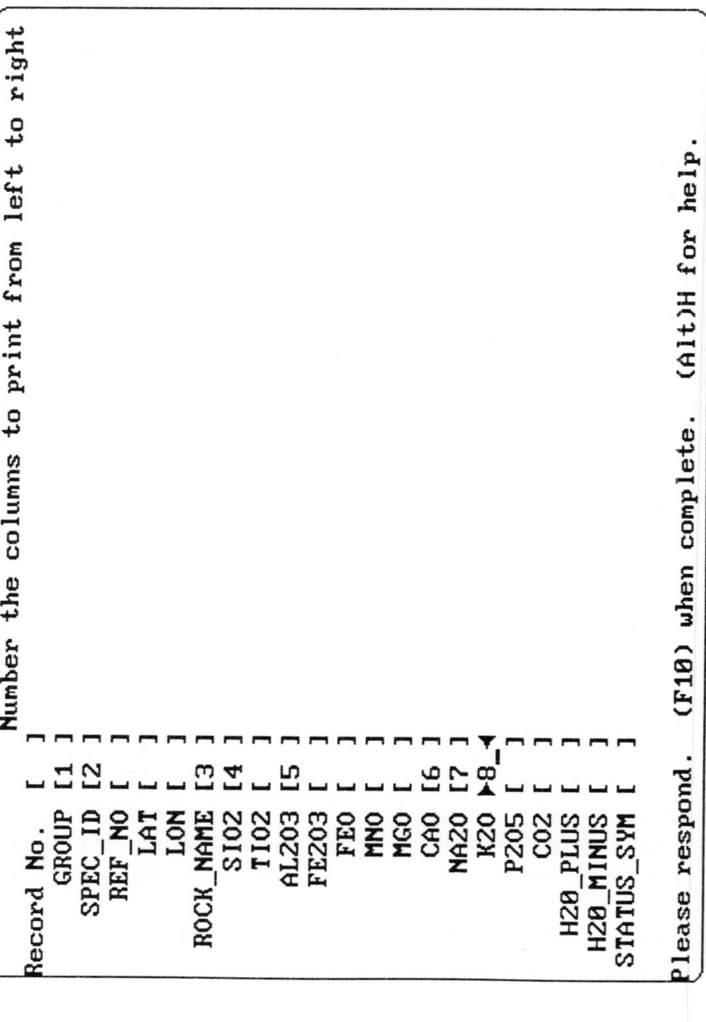

Figure 2.17 Screen for selecting fields to be used in Row Format Report.

Remember that because each record will occupy only one row, the sum of the column widths cannot exceed the width of the display (or the width of the printer).

Because a new report format is being defined, the next screen will request a "permanent title line" for the report. If the **<Enter>** key is pressed without any other response, then there will be a chance to give the report a title each time this report format is used.

The next screen is the Report Menu Screen (Fig. 2.16; the same as was used with the "page" format report). Once again, for this exercise make sure that the report is sent to the screen (press **<S>** for the option in the top row of the menu) and also press **<S>** for the option in the bottom row in order to include only selected records in the report. After these options are set, press **<F10>** to proceed, and then press **<C>** to select a Complex Search on the next screen. The Complex Search Screen (Fig. 2.12) is displayed and you must type in a "search formula" in the box at the bottom of the screen. For this example, type in ***(ROCK = "Basalt ") & (SI < 48)***. Press **<F10>** to accept the selection criteria and then when asked for a report title type in ***Sample Row Report***, followed by pressing **<Enter>**. The first page of the report should look similar to Figure 2.18. The actual entries will be the same only if you have not sorted or otherwise rearranged the sequence of records in EXTRACT. Continue pressing **<Enter>** to see the remaining two pages of records, and then press **<Enter>** once more to see the last page of the report (Fig. 2.19), which is a summary page which includes the number of selected records and the selection criteria. Press **<Enter>** to return to the Master Menu Screen.

2.9 CREATING A NEW DATABASE

Up to this point you have worked with the existing database named EXTRACT. This section will describe the steps necessary to design a new database. For the sake of simplicity, the new database will be similar to the EXTRACT database you have been using so far, but will include some extra features not available in EXTRACT. The new database will be named EXTRACT2.

```
                     Sample Row Report
June 3, 1992 at 9:59 p.m.                              Page    1

GROU SP ROCK_NAME                  SIO2   AL2O3  CAO    NA2O
==== == ========================== ====== ====== ====== ======
DJ   C  BASALT                     46.80  16.93  10.88  3.95
DJ   B  BASALT                     42.07  11.24  8.53   3.66
DJ   A  BASALT                     45.63  14.54  9.83   3.46
DL   G  BASALT                     47.40  16.00  10.00  2.60
DL   F  BASALT                     47.70  17.00  11.20  2.60
DQ   J  BASALT             ╔═ Please reply  ↑↓ ════════╗.08
DQ   I  BASALT             ║ More...  Press (Enter) ►_◄ ║.71
FM   C  BASALT             ╚════════════════════════════╝.24
FM   B  BASALT                     47.10  18.52  11.98  2.33
FM   A  BASALT                     46.00  18.24  11.56  2.34
IE   M  BASALT                     47.90  15.30  11.72  1.88
KD   B  BASALT                     46.70  13.96  9.55   2.48
KD   A  BASALT                     45.16  13.51  9.71   2.38
NB   D  BASALT                     47.01  15.57  9.77   3.00
NB   B  BASALT                     47.80  18.31  13.00  2.48
NB   A  BASALT                     47.12  14.93  9.44   2.57
NN   G  BASALT                     46.05  16.34  11.55  2.58

At 76: (((ROCK="Basalt")&(SI <48)))
```

Figure 2.18 First page of Sample Row Report.

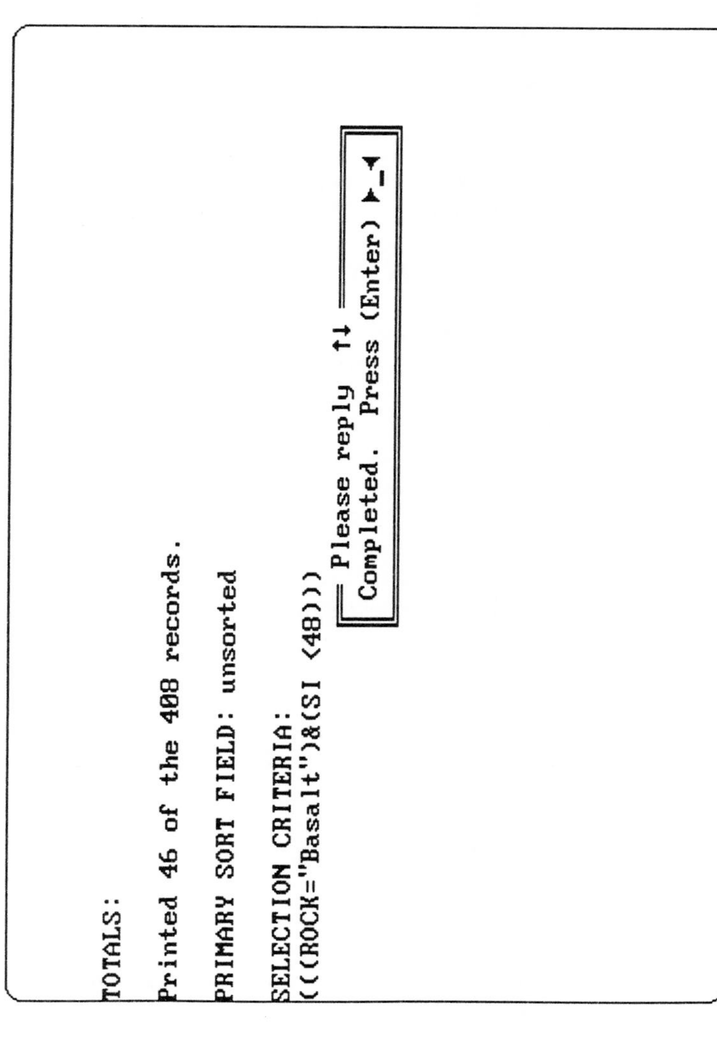

Figure 2.19 Last page of Sample Row Report.

Assuming that PC-File+ is running and that the EXTRACT database is loaded, press **<Q>** to exit from the EXTRACT database, and then press **<D>** to indicate that you want to use a different database. If you are starting from the beginning and PC-File+ is not loaded, then type ***pcf*** as before (see the section earlier in this chapter on "Loading and Starting PC-File+") to start the program and proceed as before until you reach the Database Selection Screen (Figure 2.2).

First you must give a name to the new database. The name may be up to eight characters long, and may include any valid DOS filename characters. Leave the cursor on the empty field and type ***EXTRACT2*** and press **<F10>** to accept the name. The next prompt is:

"Do you want to define it? Y/N _Y_"

Reply by pressing **<Y>** or **<Enter>**. Before a database can be created, it must be defined. The definition process involves identifying the fields which will be used to describe each record in the database. Up to a maximum of 70 fields can be used in PC-File+; the minimum number of fields is one. The description of each field must include at least a field name and a field length. As the next screen (Fig. 2.20) indicates, there are two ways to define a database in PC-File+. Press **<F>** to select the "fast" method; see the PC-File+ user method for using the "paint" method. The screen that appears has 70 blank fields into which field names may be typed. It is a good idea to create field names which are descriptive, but as short as possible. Although it is not illegal to include blank spaces in field names, it is not a good idea to do so; it would be better to include an underscore character "_" between words. In PC-File+ the last character of a field name ought to be a "#" if that field is to be a numeric field (such as a chemical analysis) and you will later want to obtain sums for that field (in reports, for instance) or use that field as part of a later calculation. In general, it is a good idea to plan ahead before you sit down to define your database, but do not worry too much about the naming and describing of fields, because in PC-File+ these descriptions are not permanent and can be changed at a later time through the options available on the Utilities Menu.

Now fill in the field names exactly as they are shown in Figure 2.21. Note the differences between the field names and locations here and those which were used in the original EXTRACT database, including the use of lower-case characters, fields located in more than one column, and

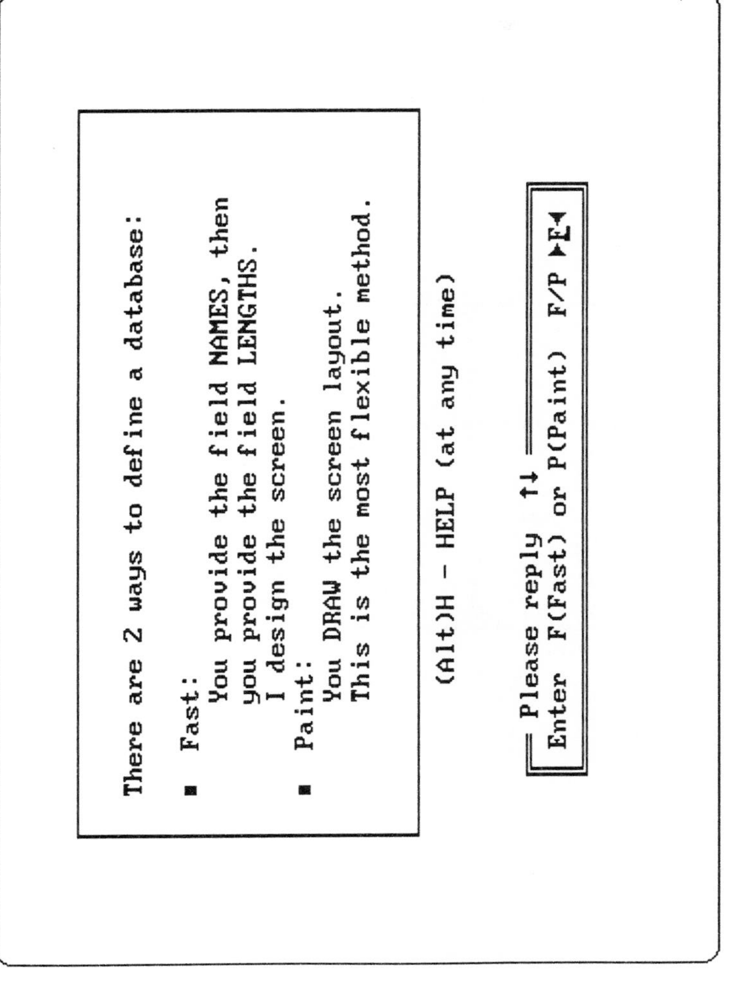

Figure 2.20 Database Definition Menu.

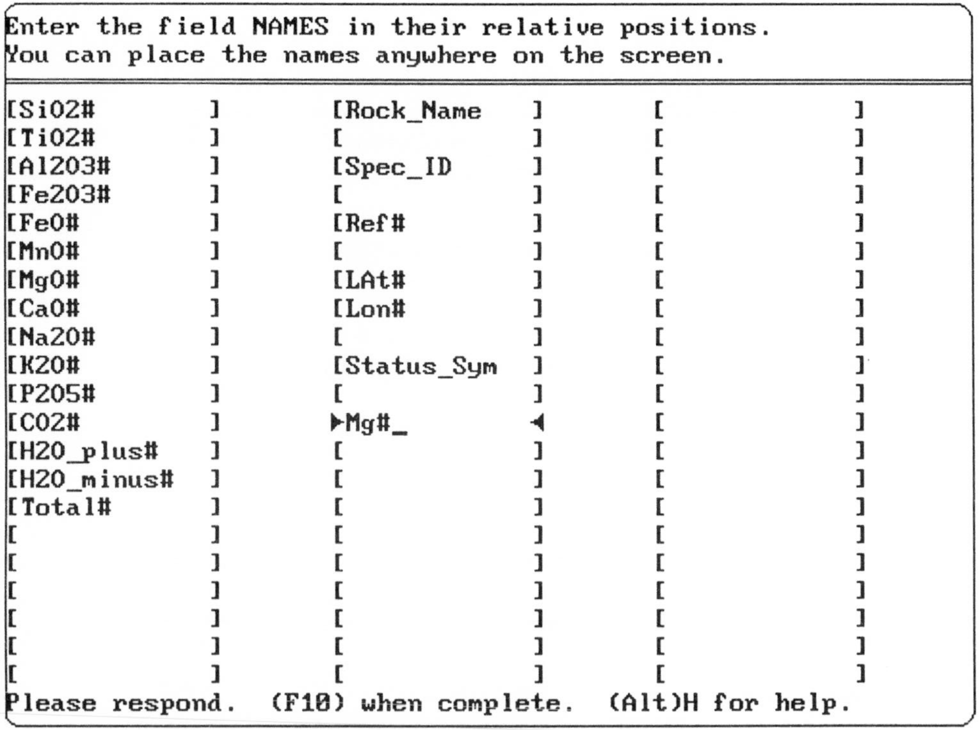

```
Enter the field NAMES in their relative positions.
You can place the names anywhere on the screen.

[SiO2#          ]      [Rock_Name    ]      [             ]
[TiO2#          ]      [             ]      [             ]
[Al2O3#         ]      [Spec_ID      ]      [             ]
[Fe2O3#         ]      [             ]      [             ]
[FeO#           ]      [Ref#         ]      [             ]
[MnO#           ]      [             ]      [             ]
[MgO#           ]      [LAt#         ]      [             ]
[CaO#           ]      [Lon#         ]      [             ]
[Na2O#          ]      [             ]      [             ]
[K2O#           ]      [Status_Sym   ]      [             ]
[P2O5#          ]      [             ]      [             ]
[CO2#           ]      ►Mg#_         ◄      [             ]
[H2O_plus#      ]      [             ]      [             ]
[H2O_minus#     ]      [             ]      [             ]
[Total#         ]      [             ]      [             ]
[               ]      [             ]      [             ]
[               ]      [             ]      [             ]
[               ]      [             ]      [             ]
[               ]      [             ]      [             ]
[               ]      [             ]      [             ]
[               ]      [             ]      [             ]
Please respond.  (F10) when complete.  (Alt)H for help.
```

Figure 2.21 Screen for naming and locating fields in new database.

the addition of two new fields (Total# and Mg#), which will be "calculated fields".

Although it is possible to define up to 70 fields for each record, having fewer fields will allow more records to be filed in the database. The field lengths defined could have been longer (up to 65 characters, or even longer if "window" fields are used), but longer field lengths would have decreased the number of records which could be filed in the database. The defined field length ought to be just long enough to accommodate the largest anticipated entry for that field. The field names can be up to twelve characters long, but it is a good idea to keep them as short as possible, because when the field names are printed above columns in row-formatted reports, if a field name is longer than the corresponding field length, then the field name is truncated. After you have named and located the EXTRACT2 fields as they are shown in Figure 2.21, press **<F10>** to continue.

Next you must define the field length for each field and the screen for this purpose looks similar to Figure 2.22. Fill in the defined field lengths exactly as shown in Figure 2.22 and then press **<F10>** to continue. The next screen asks if any of the defined fields in the new database are "window fields". Window fields are fields for which the display length (length shown on the screen) is less than the defined length of the field. The content of these fields is scrollable within the displayed length. Since none of the newly defined fields are window fields, press **<N>** to continue.

The next screen allows you to control the sequence in which a record's fields are accessed. Fill in the screen exactly as shown in Figure 2.23 and press **<F10>** to continue.

PC-File+ then will create the new EXTRACT2 database files and return you to the initial Master Menu Screen (similar to Fig. 2.3). But before you can add any new records to EXTRACT2, it is necessary to define the Total# and Mg# fields. This is accomplished by going to the Utilities Menu. From the Master Menu Screen press **<U>** or **<F8>** and the PC-File+ Utilities Menu (Fig. 2.4) will appear. Then press **<N>** to indicate you want to change (in this case, add) a field calculation.

The next screen lists all of the EXTRACT2 fields and asks which field you want to change. Type ***15*** in the space provided in order to indicate you want to change the Total# field first. The next prompt occurs in a box across the middle of the screen:

"Change (N)Name (M)Mask (K)Constant or (C)Calcs _N_"

```
Enter the display length (1 - 65) for each field

SiO2#         [6 ]    Rock_Name    [25]
TiO2#         [6 ]
Al2O3#        [6 ]    Spec_ID      [6 ]
Fe2O3#        [6 ]
FeO#          [6 ]    Ref#         [5 ]
MnO#          [6 ]
MgO#          [6 ]    LAt#         [9 ]
CaO#          [6 ]    Lon#         [9 ]
Na2O#         [6 ]
K2O#          [6 ]    Status_Sym   [40]
P2O5#         [6 ]
CO2#          [6 ]    Mg#          ►6_◄
H2O_plus#     [6 ]
H2O_minus#    [6 ]
Total#        [6 ]

Please respond.  (F10) when complete.  (Alt)H for help.
```

Figure 2.22 Screen for defining field lengths in new database.

```
          Please number the fields in the order desired

      SiO2# ►1 ◄        Rock_Name [16]
      TiO2# [2 ]
     Al2O3# [3 ]          Spec_ID [17]
    Fe2O3# [4 ]
       FeO# [5 ]             Ref# [18]
       MnO# [6 ]
       MgO# [7 ]             LAt# [19]
       CaO# [8 ]             Lon# [20]
      Na2O# [9 ]
       K2O# [10]       Status_Sym [21]
      P2O5# [11]
       CO2# [12]              Mg# [22]
  H2O_plus# [13]
 H2O_minus# [14]
     Total# [15]

Please respond.  (F10) when complete.  (Alt)H for help.
```

Figure 2.23 Screen for defining sequence in which fields will be accessed in a new database.

Press **<C>** to indicate you want to change the Total# field calculation. Then another prompt box is displayed across the middle of the screen. Type the calculation formula into the box exactly as it is shown in Figure 2.24. Note that the formula shows the sum of the oxide fields and that it is necessary only to type enough of each field name to identify that field uniquely. Press **<Enter>** after the formula has been entered.

Next, repeat the process by typing **22** (followed by pressing **<F10>**) in the space at the top right of the screen in order to add a calculation of the "magnesium number" (field Mg#) for each record. Then press **<C>** to select a calculation change and type the following formula into the prompt box exactly as shown next:

(MgO/(MgO+FeO+Fe2))

Press **<Enter>** to accept this calculation. This will return you to the Utilities Menu. Press **<Q>** to return to the Master Menu Screen.

Now you are ready to add records to EXTRACT2. Follow the directions for adding a record which were given earlier in this chapter. After you type your data into the fields for each record and press **<F10>** to accept it, you will note that the contents of the Total# and Mg# fields are calculated automatically. When you are adding records to a PC-File+ database, remember that use of the **<Ctrl><D>** and **<Ctrl><F>** duplication functions can be useful (see Table 2.2).

2.10 OTHER PC-FILE+ FUNCTIONS

2.10.1 Creating Graphs

The PC-File+ graphing functions and options are not useful for working with geological databases. You are restricted by the program to plotting sums (PC-File+ sums the contents of the specified field(s) for the records in the database) versus categories (e. g. rock names). You can not create scatterplots easily for two variables (fields) in which each record is represented on the graph by a single point.

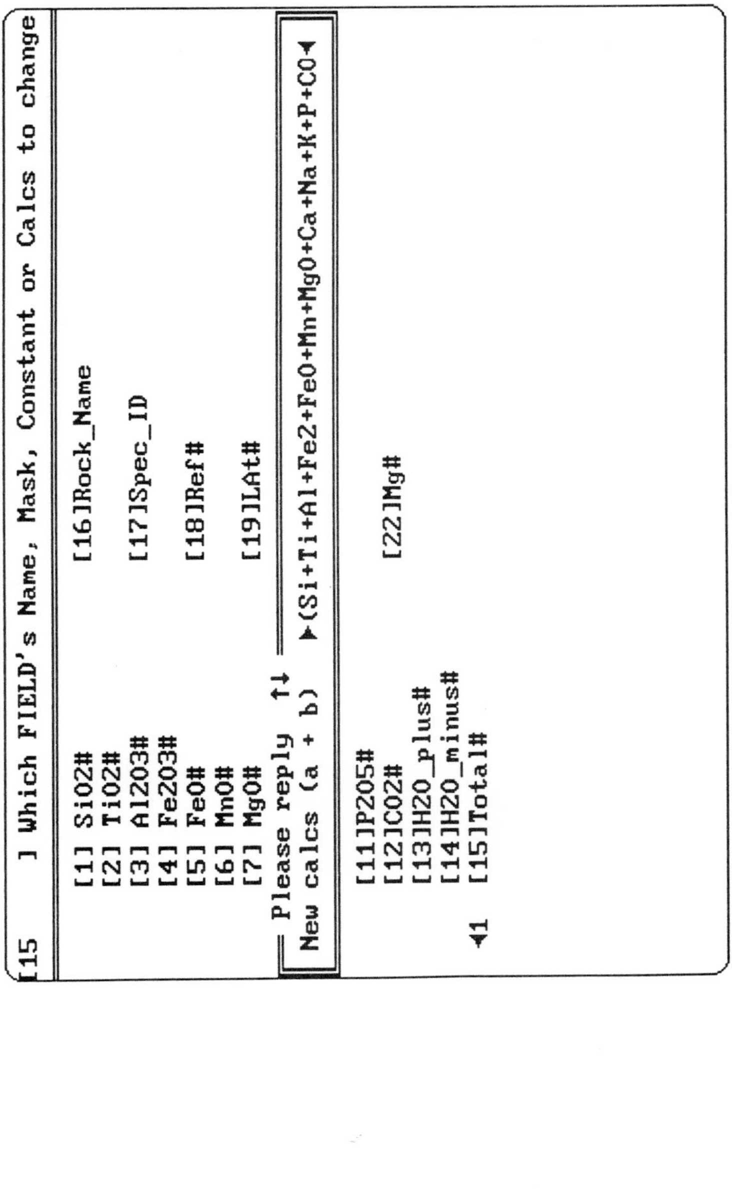

```
[15      ] Which FIELD's Name, Mask, Constant or Calcs to change
     [1] SiO2#              [16]Rock_Name
     [2] TiO2#
     [3] Al2O3#             [17]Spec_ID
     [4] Fe2O3#
     [5] FeO#               [18]Ref#
     [6] MnO#
     [7] MgO#               [19]LAt#
 ═ Please reply  ↑↓ ═══════════════════════════════════
  New calcs (a + b)   ►(Si+Ti+Al+Fe2+FeO+Mn+MgO+Ca+Na+K+P+CO◄

     [11]P2O5#
     [12]CO2#               [22]Mg#
     [13]H2O_plus#
     [14]H2O_minus#
 ◄1  [15]Total#
```

Figure 2.24 Screen for defining calculated fields in new database.

2.10.2 Utilities

The PC-File+ Utilities Menu offers a wide variety of database utility functions, some of which were used during the exercises in this chapter. Databases can be imported and exported in a variety of popular formats. PC-File+ database (and other types of) files can be copied, deleted, or re-named. Changes can be made in field names, field masks (which limit or format data input into fields), field constants, and field calculations. The current PC-File+ database can be "cloned", while adding, deleting, or modifying field names and lengths at the same time. Duplicate records in the same database can be located. And, under certain circumstances,, records deleted previously can be "undeleted".

You are encouraged to try using these and other PC-File+ functions not described in this chapter. The PC-File+ user manual and the context-sensitive help screens will help you.

CHAPTER 3

An Introduction to dBASE Database Management Programs

Joseph Frizado
Dept. of Geology
Bowling Green State University
Bowling Green, Ohio USA

3.1 INTRODUCTION

The dBASE series of database management programs is one of the most widely used software packages for managing data with a micro-computer. The primary ancestor of the dBASE series of programs was named VULCAN and was written in the late 1970s for a CPM-based microcomputer by Wayne Ratliff. With the advent of the IBM-PC, it was upgraded into dBASE II by Ratliff and George Tate (Jones,1987). dBASE II's strongest point was its BASIC-like interpreted language. However, this also was its weakest point as a user had to learn a new computer language in order to use the software. Because of this situation, many people learned dBASE II and wrote application programs for those who did not wish to learn the language. Many specific business applications were written in the dBASE language, but only a few programs had any scientific use.

As the market for database management systems grew, other companies developed programs to augment some of dBASE IIs shortcomings. Because the dBASE language is interpreted, it operates slowly. Pseudocompilers, such as Clipper (Dinerstein, 1987), were written in order to increase the speed of dBASE applications. The native language of Clipper basically was the same as dBASE, but it has additional capabilities and can be compiled into executable code. Programs to facilitate generating dBASE code also were created such as Genifer. This type of software allowed a user to select items from a menu and have the application compose dBASE code to complete a finished dBASE program. As dBASE II aged, other DBMS systems improved as did the accessory programs of other developers, so the Ashton-Tate released an update, dBASE III. This version of dBASE was only an extension of dBASE II with an increase in the size and number of files that the program could handle. Additional mathematical functions were added and a vastly improved sorting routine was implemented. New commands were added to the language, but the system was extremely complex and cumbersome to the novice user. dBASE III also was known to be a "buggy" system. In 1985, dBASE III Plus was created by fixing the bugs in dBASE III and adding two new features, the Assistant and the Applications Generator. It was at this point that usage of dBASE programs increased dramatically.

As with all DBMS packages, dBASE III Plus assumes that you already know how to use your computer and that you understand the rudiments of DOS. The software itself is a database management language as was its predecessors. The fact the dBASE really is a language and not a menu-driven package always has been both its strongest and its weakest point. For the novice, the package can be overwhelming and difficult to learn quickly. However, the advanced programmer can manipulate a database easily using the dBASE language capabilities to create menu-driven programs for other users. When dBASE III Plus was released, the manufacturers decided to develop methods to ease the novice user into the system by two pathways, the Applications Generator and the Assistant. The Applications Generator is a large program which can write simpler programs in the dBASE command language to perform some operations. The user runs the Applications Generator, selects items from a series of menus and the generator creates an application (or program) to perform that operation. The Assistant is a menu-driven system that can be invoked with the dBASE command **assist**. When the dBASE III Plus program is first executed, it enters the Assistant automatically. An experienced user can disengage this feature and use the command language directly. The Assistant is a useful teaching tool to learn the rudiments of the dBASE language. However, it is powerful enough to perform many of the data processing operations of interest to a geologist and may be all that some users may need initially.

3.1.1 dBASE III Plus and dBASE IV

In 1989, Ashton-Tate upgraded the dBASE series of programs with the release of dBASE IV. Although there is a large installed base of dBASE users, there has not been a great rush to upgrade to the new version. The reasons for this are two-fold: (1) experienced users already have the additional features of dBASE IV via other software, and (2) users do not feel the extra features are worth the upgrade price. Because of this, there are quite a few copies of dBASE III Plus that will not be upgraded. In this chapter, features present in both versions of dBASE as well as new features found only in dBASE IV will be discussed.

dBASE IV as an upgrade to dBASE III Plus adds a series of features (Table 3.1) and requires more extensive hardware. dBASE III Plus can be used on a machine with as little as 256K of memory. dBASE IV requires at least 640K and a hard disk drive. These requirements are a minimal change since many IBM compatible microcomputers have at least 640K and if the user is serious about having a database of any size, a hard disk drive already is part of their microcomputer system.

Additional functionality to the programming area is the main advantage of upgrading to dBASE IV. Of special note are the Control Center, the addition of a SQL interface, and the addition of a compiler. In addition to technical enhancements, Ashton-Tate also has revised the menu interface of dBASE III Plus. The Assistant was a program that came with dBASE III Plus to give the novice user a menu-driven interface into database operations. Each menu item selected leads the user through a series of options to perform an operation. For example, if you select Database File under the Create menu item, the program will ask you where the file should be placed as well as what the name will be before it creates the database. Many of the operations that can be performed directly from the language interface can be selected via the Assistant. As selections are

Table 3.1 Additional features of dBASE IV

Control Center - replaces Assistant with activity-oriented menu system.

Reports, Forms, and Labels - these formats are stored in binary files. If companion files are missing, needed files are generated by dBASE IV. There is more flexibility with each as they now can access multiple databases, full editing capabilities, and in general a much improved report generating facility.

Multiuser security - on multiuser network, locked databases can indicate who locked them, accessory files can be locked at the same time as database, and multiple users can access same record at same time.

Queries - dBASE IV offers both Query by Example and SQL commands for linkages to mainframe databases.

Memo Fields - in dBASE IV, memo field can be 64,000 characters rather than dBASE III Plus' 47,000.

Compiler - dBASE IV has compiler and it is invisible to user. Programs that can be compiled are compiled automatically before being run.

Multiple relations - under dBASE III Plus parent database could have only one child.

Every motion in databases was linked. This can be bypassed under dBASE IV.

New indexing - Master File Index can hold 43 indexes on database.

New text editor - dBASE IV has improved text editor.

New commands - WINDOW, RESTORE WINDOW FROM, user definable commands and others.

made, the Assistant displays the command line equivalent of the operation near the bottom of the screen. By noticing what the program displays in this area, the novice user can learn about the dBASE language. Each selection in the Assistant fills in the blanks for a dBASE command line operation. When the user is done making selections, the Assistant runs the single command line it has built. In dBASE IV, the Assistant was replaced with the Command Center. Philosophically, the main difference between the two is that the Control Center is action oriented whereas the Assistant builds each command in full view of the user. The Command Center also is menu driven, but more complex than the Assistant. More operations can be performed from the Command Center. This is achieved by the use of more pop-up menus and hierarchical menus rather than the simple menu pattern of the Assistant. Each of the menu-driven systems has its advantages and both lead you to the same sequence of screens on many operations.

3.2 THE ASSISTANT

When you start dBASE III Plus the computer will load the various overlays and configuration files needed before showing you a screen that contains the license agreement. After accepting the license agreement terms, dBASE II Plus will default into the Assistant (see Fig. 3.1). The Assistant screen is divided into three sections (from top to bottom), the menu line, the main body of the screen, and a status area at the bottom of the screen as in Figure 3.2. The menu line consists of a series of words. When one word is selected, the program will allow you to proceed with any commands grouped under that heading (see Fig. 3.3). You may select a major heading by moving either the cursor to that selection with the directional keys or by entering the first letter of the heading you wish to select. After you select a heading, an expanded menu of the next layer of options will be displayed extending into the main body of the screen as a pull-down menu. Selection within the pull-down menu is done by using the directional cursor keys. By pressing **<enter>** you will select the highlighted item. If that selection has options, they will be presented to you in an accessory box in the main body of the screen.

The main body of the screen is utilized only for display of options from the selected commands, possible selections for some options and

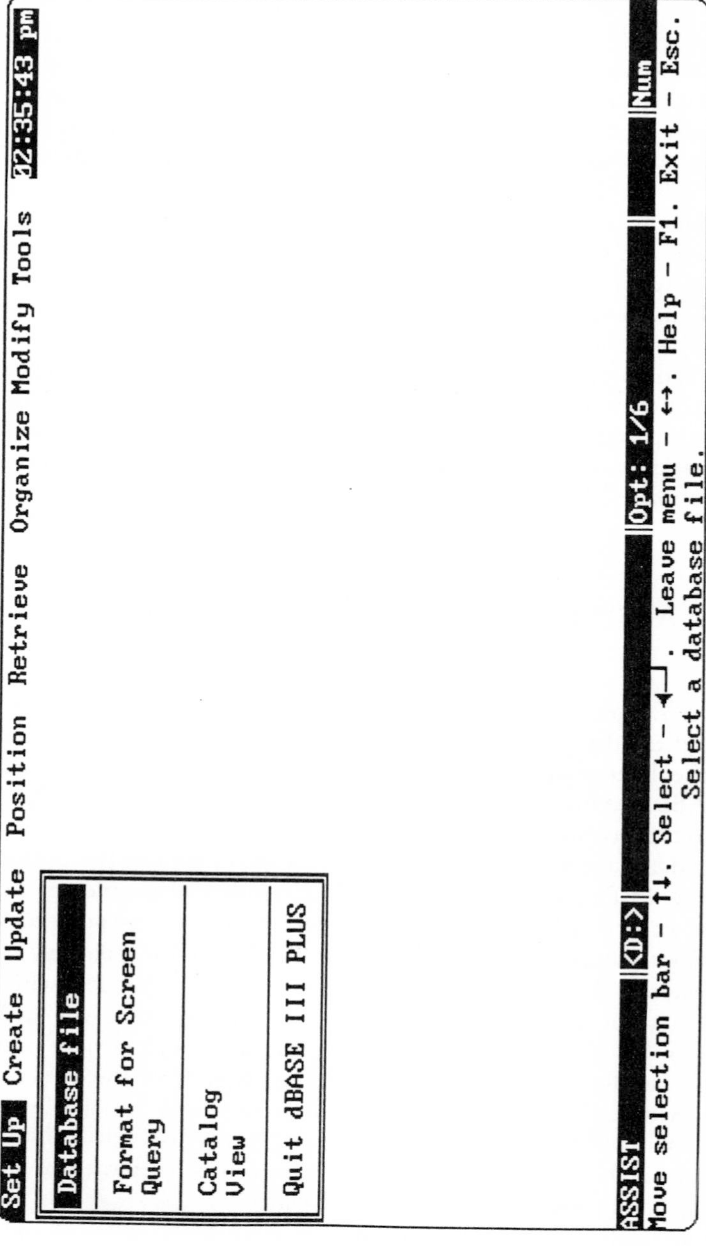

Figure 3.1 Assistant main screen. After starting dBASE III Plus this is what the user sees.

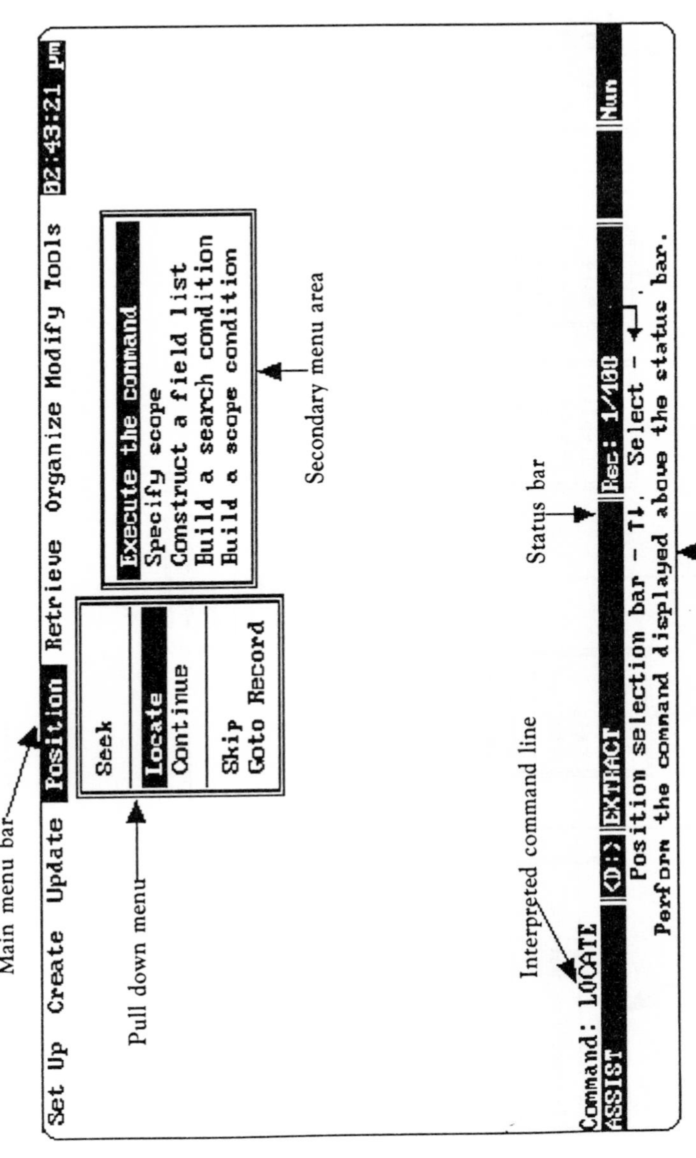

Figure 3.2 Functional areas of the Assistant screen.

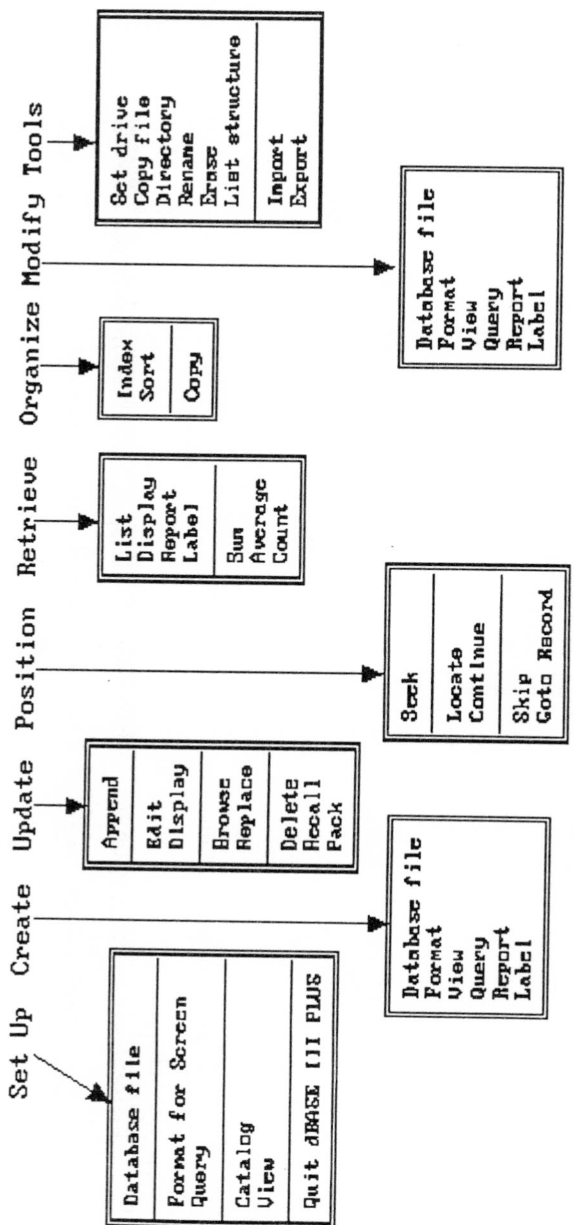

Figure 3.3 Options under main menu of Assistant. As you select any item on main menu bar,pull-down menu with options is displayed below main menu bar.

data. In the first two circumstances, the user cannot change what is displayed. You may only move the cursor and select the pertinent item(s).When the screen is being used to display data from the database itself, you can edit the data displayed on the screen if you are in one of the edit modes.

The status area at the bottom of the screen is composed of three items: (1) one or more lines explaining options available from your current position; (2) the status line itself; and (3) the dBASE language command being created by the Assistant. Explanatory text on the bottom of the screen is helpful when used in conjunction with the selections made via the menus. The status line displays information about the database that you are using currently, where you are within that base (record position) and what type of mode you are in (insert, caps lock, numerical lock, etc.). In using the Assistant, you should begin to develop your skills in the dBASE III Plus language by noting that the relevant dBASE III Plus command is displayed on the line above the status line at the bottom of the screen. As you make selections, the Assistant will fill in the various parts of a complete command line to perform the operation that you are requesting. However, the Assistant utilizes only some of the dBASE III Plus commands and only some of the options within those commands. The dBASE III Plus language is much more powerful and flexible than suggested by using the Assistant.

If you are using the Assistant to perform the exercises, please keep in mind the following points:

(1) To move around the Assistant main menu screen you can sometimes press the key of the first letter of the item that you wish to choose, but you always can use the cursor keys to navigate between and within the various menus, to highlight your choices, and make your selections.

(2) To back out of a process or step, press **<esc>**. This will take you back one step towards the main menu. However, if you press **<esc>** one time too many, you will leave the Assistant and see the dot prompt "." on the lower left of the screen. At this point you are in the programming language editor and can enter single commands to have dBASE III Plus perform a particular process. If you wish to reenter the Assistant, which is what you want to do for most exercises, please type in ***assist*** and press **<enter>**. Pressing **<F2>** also will start the Assistant from the dot prompt.

(3) The Assistant allows you to make particular choices about the operation you are performing. It only allows you to select usual options that most people wish to use most of the time. The dBASE III Plus programming language has many other options that are not accessible from the Assistant.

(4) The status bar (see Fig. 3.2) on the bottom of the screen contains information about your current situation. It displays where the database is stored and its name. It displays your current position in the base, the number of records in the entire base, and if you have activated the caps lock or num lock keys on the keyboard. It also tells you if you are in the insert or normal (delete) mode. Hints- if you ca not get the correct case for any entry look at the status bar to see if "caps" is activated. If the cursor arrows are inoperable in selecting items on a menu, look at the status bar to see if "num" is activated.

3.3 THE COMMAND CENTER OF DBASE IV

One of the perceived flaws of dBASE III Plus is that the Assistant is not as useful or versatile as it could be. People using the Assistant tend to like the interface but are unable to access all of the features of dBASE III Plus. In essence, users wanted an operational interface, one that would perform a given operation rather than build a command line. The Command Center of dBASE IV answered these criticisms by adding many more features and by creating an operational menu system rather than using the Assistant's approach. Although this makes the Command Center more powerful, it can be confusing to the novice and is not as good a "teaching tool" as the Assistant.

When you start dBASE IV, the computer will spend some time loading the various overlays and files that the program requires before displaying the license agreement and the main screen of the Control Center (see Fig. 3.4). As with the Assistant, the Control Center screen can be divided into three sections. The top menu bar of the screen allows the user to change system options or move to DOS through **Catalog, Tools** and **Exit.** The central portion of the screen is a second series of menus that shows you what is currently contained in the default catalog. The six headings, **Data,**

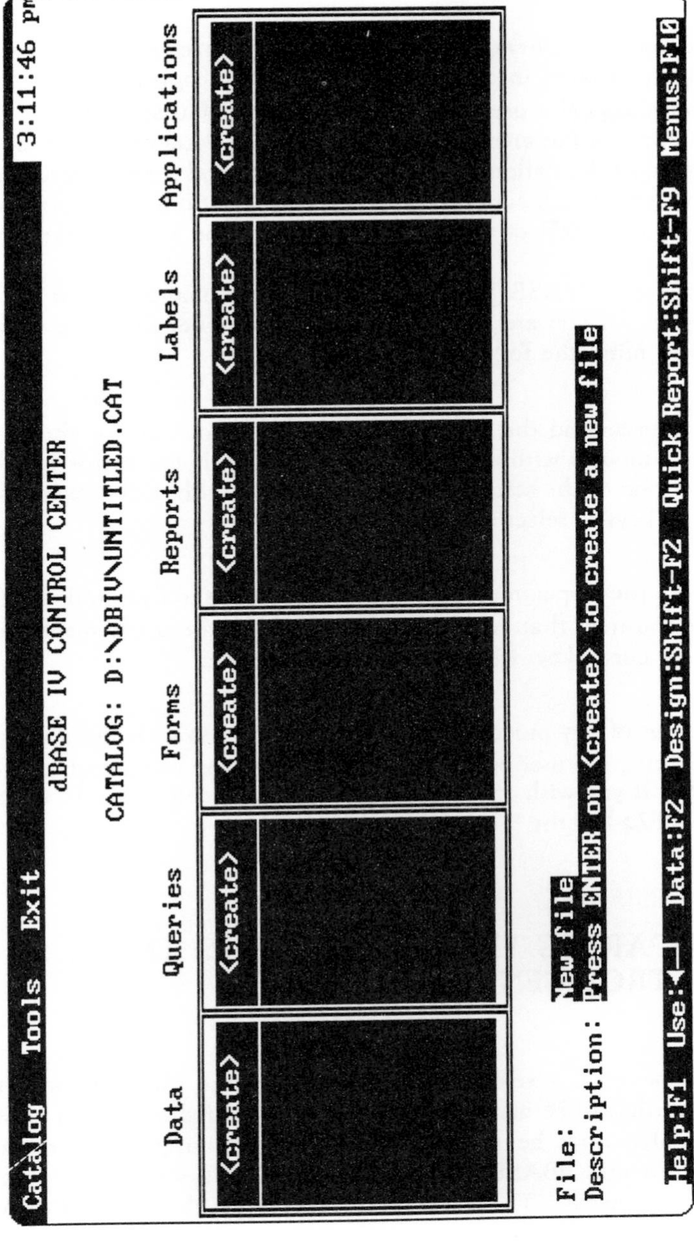

Figure 3.4 Control Center main screen. After starting dBASE IV thsi is what the user sees.

Queries, Forms, Reports, Labels, and **Applications** correspond to different file types used in dBASE IV. The **create** option at the top of each column allows the user to create a new file of that particular type rather than access a file created previously. The lowermost portion of the screen gives you information about where you are and a short description of the option highlighted currently . Finally, the notations at the bottom of the screen are reminders of what some of the function keys can do at this particular juncture.

Unless altered, dBASE IV will enter the Command Center as the program begins. If you are using the Command Center for our exercises, please keep in mind the following points:

(1) To navigate around the Control Center screen you can use the cursor keys (or a mouse) within a selected menu area. If the cursor is in the central region of the screen, then you can move within that region with the cursor keys to select an option.

(2) To access the upper menu bar of options, press **<alt>** and the first letter of the item that you wish to select. Once the upper menu bar is active, the cursor keys will work within this region.

(3) To back out of any process, use the **<esc>** key. As with the Assistant, this will bring the user back out one level until you have reached the ". " prompt. If you wish to restart the Command Center, type in ***assist*** (or press **<F2>**) at the "." prompt.

3.4 COMPARING THE ASSISTANT TO THE CONTROL CENTER

For some novices, a selection between using the Assistant and the Control Center has to be made. Although each is mated to its own version of dBASE and cannot be interchanged, anyone having access to both dBASE III Plus and dBASE IV will have to decide which approach is optimal for their needs. The selection depends upon their uses of dBASE and what they feel to be important.

The Assistant has strong points in its favor. It is easier to use than the Control Center and therefore more attractive to the computer novice. It is a better learning tool than the Control Center. Within the Assistant, each action helps build a dBASE command that is displayed at the bottom of the central portion of the screen. By paying careful attention to this area, the user can learn the rudiments of the dBASE language by using the Assistant. As a user becomes more proficient in using dBASE, the ability to use the dot prompt directly and also compose programs becomes more important.

The Control Center was created to provide the user with a more powerful menu-driven interface to dBASE. Because the Assistant has only a limited selection of options, many users wanted the newer menu system to be more versatile and not necessarily provide the learning interface of the Assistant. The Control Center fulfills that requirement. It is more difficult to use than the Assistant for the novice user, but provides the advanced user with more features of the "." prompt command line. These additional features are what makes the Control Center more difficult for the novice to learn.

If the choice exists, that is you have both versions of dBASE, use the Assistant for the casual user who knows little about computers. Also, if you wish to learn the rudiments of the dBASE language, the Assistant is the better selection for getting started. The Control Center is for those users who wish to perform more intricate manipulations, but do not wish to learn the command line language of dBASE. In either situation, use these facilities to become familiar with dBASE. If you have access to both, please remember that the command line language of dBASE III Plus is appropriate for dBASE IV, so you may use the Assistant to learn dBASE with every intention of migrating to dBASE IV at a later date. Regardless of your final choice of menu systems (and whether you wish to use the dBASE III Plus or dBASE IV), eventually you will probably work directly on files from the "." prompt.

3.5 FILE NOMENCLATURE IN DBASE

As you begin to work with dBASE, you will inevitably need to decipher the file nomenclature used by the program. Filenames under MS-DOS can have up to eight characters followed by a period and a three letter extension. A variety of extensions is used to identify the different types of files within dBASE. Some of the more usual extensions are:

.BAK a back-up file. This file can be a duplicate of a database file, procedure file, or program file.

.BIN a binary file.

.CAT a catalog file. This file contains a selection of files that the user has selected to group together.

.DBF a database file. This file contains the data.

.FMT a format file. This file contains information about how the data will be viewed on a CRT screen.

.FRM a report format file. This file contains the format for printing out a table of the data from the database.

.NDX an index file. This is the file that contains index information for a database.

.QRY a query file. This file stores a particular query or filter to be applied to a database for record selection.

.SCR a screen file. This file describes a screen format for the display of data on the CRT. It is a companion file to .FMT.

.TXT a text file. This is an ASCII text file to be read by other software.

.VUE a view file. This file stores a description of the relationships between linked databases that together comprise a view.

.PRG a program file. This text file is a user created program of dBASE commands.

For dBASE IV users, additional files may be present, as well as compiled version of some files:

.DBO a compiled program. This is the compiled version of a .PRG file.

.FMO a compiled screen format. This is the compiled version of a .FMT file.

.FRG a report format. This file contains all of the formatting information to generate a user created report.

.FRO a compiled report format. This is the compiled version of a .FRG file.

.LBG a label format file. This file contains all of the formatting information to print a series of mailing labels.

.LBO a compiled label file. This is the compiled version of a .LBG file.

.QBE a query by example file. This file contains the information for recreating a query by example.

.QBO a compiled query by example file. This is the compiled version of a .QBE file.

.UPD a update query file. This file contains all of the changes made to a .QBE file.

.PRS -a SQL program file. This file contains the SQL commands used to access a SQL type database.

The best way to understand the purpose of each file type under the dBASE systems is by example.

3.6 USING A PREEXISTING DATABASE

If the database that you wish to use is in dBASE format, you select it from the **Set Up** menu of the Assistant or select it from the **Data** column of the Control Center. You will need to know the path to the file, but both programs will display a list of .DBF files in the current directory. The user selects a database from the list by positioning the cursor and selecting that entry. If your are using the Assistant, please select the EXTRACT database by pressing **<enter>** with **Database file** highlighted under **Set Up** and then by selecting a path to get to the file EXTRACT.DBF. dBASE III Plus will then ask you if the file is indexed. An indexed database has an associated file that records the positions of records based upon the values in a key field selected by the user. In the Assistant, as you answer each question, notice that in the lower left-hand screen area, a dBASE command (preceded by **Command:**) appears. This single command ***USE <database name>*** can be used to select a database for use under dBASE. If an index file exists and you wish to use it, an additional phrase ***INDEX<index filename>*** can be invoked as part of the USE command. An index for a database can be created after you have selected the database by selecting **Index** from the pull-down menu of the **Organize** entry on the Assistant's main menu. After selecting the EXTRACT database, the status line should list both the database as well as the number of records and current position of the pointer. In this situation it should read **None**.

In dBASE IV, selecting a database file from the Control Center is easy, once that file is contained in the catalog that you are using. Catalog files allow you to associate several files together into a single system. Under the **Catalog** item on the main menu, you will locate the **Add file to catalog** command. Select this command to display a menu of the files in the current directory that could be added to the present catalog. Select the EXTRACT.DBF file and it will now be displayed in the **Data** column of the central screen display. Move the highlight bar to the **Data** column in the central area of the screen. Use the cursor arrows to move from **create** to **EXTRACT**. By pressing **<enter>** when **EXTRACT** is highlighted, you will be selecting the database. A secondary screen will appear with three options. In this example, we will only be using the file, so select **Use file** as your option and press **<enter>** to activate it. At this point, on the status line you should see that you are in the EXTRACT.DBF database. However, there is no information stored in the database at this time.

3.7 FIELDS

Fields are assigned areas in each record of the database into which data are entered. If your database were of rock chemistries, then it would be appropriate to design a single form that contains marked spaces for each element analyzed. If your data set were only major element chemistries of rocks, then you would need spaces for SiO_2, Al_2O_3, TiO_2, etc. regardless if TiO_2 was not analyzed in each rock. Each element would be a field within a particular rock's data record analogous to spaces on a form to be filled out. You may name these fields anything that you wish without using blanks. However, it is convenient to use easily recognizable field names, such as SIO2 for SiO_2. Once named, the program will recognize that this particular entry is termed SIO2. That does not mean that the program can interpret a data set and find that SiO_2 is read in as the fourth number and not the third. If the third field is named SIO2 by the user, it is the user's responsibility to be sure that the third entry of each record contains the correct value for SiO_2 stored in SIO2. Once the fields are defined, they may be placed in particular positions on screen and edited to create a customized form. A form created previously (FORM1) is contained on the accompanying disk for the dataset that we will be using as an example.

Each of the fields in a preexisting database was defined when that database was created. Databases under dBASE III Plus can have to 128 fields, while under dBASE IV a database can contain 255 fields. Each field can be defined as being logical (true or false), character (A-Z,a-z,0-

9), numeric (0-9 with or without a sign), date (month/day/year), or memo (only one memo field per record with data in this field treated as characters). dBASE IV has an additional numeric filed type "F" for floating point numbers. Under dBASE IV, numeric fields ("N") are stored as binary coded decimal numbers while "F" fields are stored as floating point numbers. The only difference is how the values are stored, which affects the ultimate accuracy of any operations performed using these fields. The numeric field can accommodate up to 19 bytes. When comparing numerical values, accuracy is carried out to 13 digits. The memo field is useful if you have a character string that exceeds the maximum size (254 characters under dBASE III Plus, 1024 characters under dBASE IV) for a character field. Each memo field has a maximum length of 5000 characters and is stored as a separate file. In many situations, the user can discern what must be the type of data in each field, but it is not always obvious what the originator of the database had in mind. In order to know what the structure of the database is, use the ***LIST STRUCTURE*** command or select **List structure** from the **Tools** menu of the Assistant or the **Data** menu of the Control Center (select **Modify structure/order** in the secondary item box). Specifications of all of the fields in the database can be listed to a printer for a permanent copy. As a first step, let us examine the EXTRACT database. The ***LIST STRUCTURE*** command shows you how each field is defined and its field name in the database (see Fig. 3.5). Having determined the EXTRACT structure, let us look at the actual data.

3.8 IMPORTING AND EXPORTING DATA

Unfortunately, dBASE like most DBMS systems written for microcomputers, was written primarily for business applications. In business, most of the data entry is not from a machine readable source, but more likely entered by hand. Hence the easiest data entry technique for dBASE is via the keyboard. The user designs a form and enters sequentially data into each field until a record is completed. A new record is displayed automatically and the data entry continues. In geology, there are other methods of data entry that are more useful.

In both dBASEs, the user can select the import (or export) function for providing ties to other microcomputer applications. The options for import

```
 Set Up  Create  Update  Position  Retrieve  Organize Modify Tools  02:57:04 pm
Structure for database: D:EXTRACT.dbf
Number of data records:      408
Date of last update   : 06/03/92
Field  Field Name  Type        Width    Dec
    1  GROUP       Character       4
    2  SPECIMEN    Character       2
    3  REFNO       Numeric         4
    4  LATITUDE    Numeric         7      3
    5  LONGITUDE   Numeric         8      3
    6  ROCKNAME    Character      23
    7  SIO2        Numeric         5      2
    8  TIO2        Numeric         5      2
    9  AL2O3       Numeric         5      2
   10  FE2O3       Numeric         5      2
   11  FEO         Numeric         5      2
   12  MNO         Numeric         5      2
   13  MGO         Numeric         5      2
   14  CAO         Numeric         5      2
   15  NA2O        Numeric         5      2
   16  K2O         Numeric         5      2
Press any key to continue...
ASSIST       ||<D:>||EXTRACT          ||Rec: 1/408        ||     ||Num
```

Figure 3.5 List Structure under dBASE III Plus. Results under dBASE IV are similar.

in either situation are limited to a few program formats. It is usual to use a more universal file format such as a tab-delimited or comma-delimited file. In each of these files, entries are separated by a tab or comma, with character strings enclosed in quotes. This type of file usually can be exported or imported into a variety of programs. In our example, we are interested in adding information to our EXTRACT database from a comma-delimited file named EXTRACT.CDF.

The most flexible method for adding data to our database is via the append command entered at the dot prompt outside of the Assistant and the Control Center. Press **<esc>** until you have exited the menuing system and the screen resembles Figure 3.6. At this point you could enter any of the possible forms of append:

Append from <filename> type sdf- appends from a file in System Data Format file.

Append from <filename> type dif- appends from a Data Interchange Format file.

Append from <filename> type sylk- appends from a Multiplan spreadsheet in row order format.

Append from <filename> type wks- appends from a spreadsheet (Lotus) format file.

Append from <filename> type delimited- appends from a file with fields that are separated by a comma.

Append from <filename> type delimited with blank- appends from a file with fields that are separated by a blank character.

Append from <filename> type delimited with <char> - appends from a file with fields that are delimited by a user-specified character.

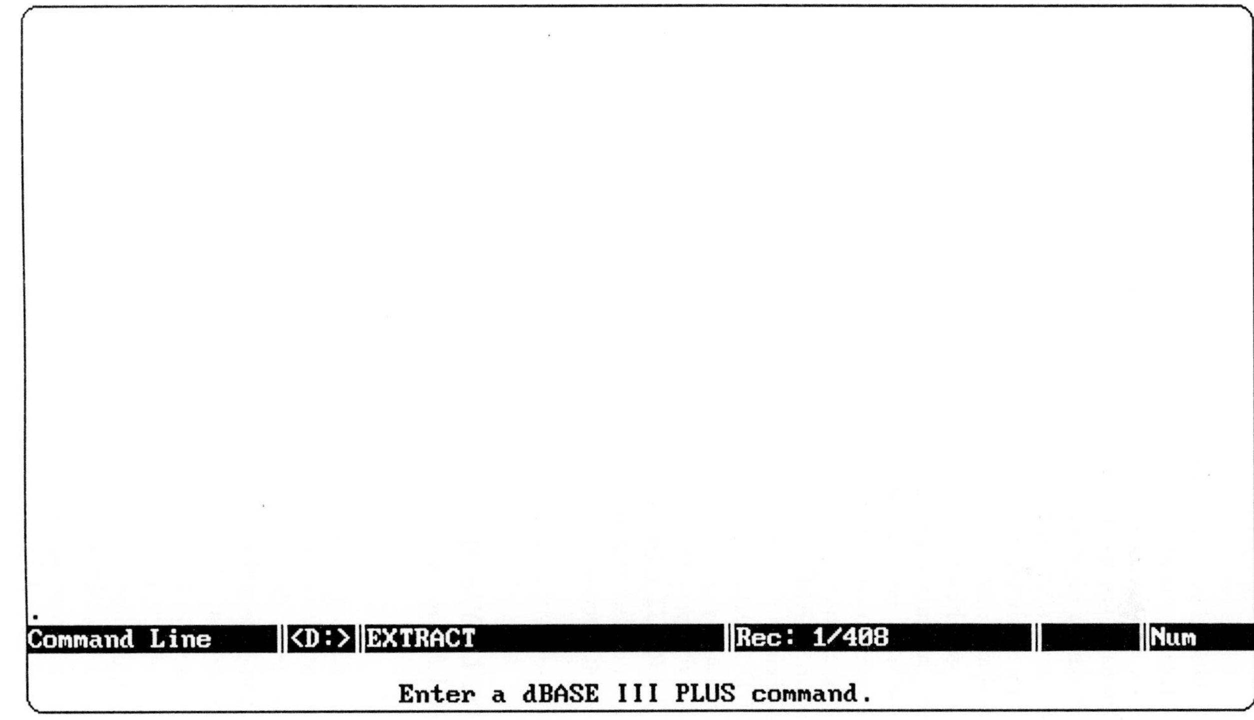

Figure 3.6 Dot prompt (".") screen of dBASE.

In each situation, you should know the structure and field names in the incoming data, so that you can create an appropriate dBASE III Plus database. Exporting data to other programs can be accomplished via the ***Export*** command (to specified formats) and via the ***Copy*** command. ***Copy*** allows the user to copy the existing database as another file in the same variety of formats available under Append and can therefore export a file into a spreadsheet, graphics, or other program.

The database used as an example in this chapter is the EXTRACT database of geochemical information on 408 igneous rocks described earlier (see Chapters 1 and 2). In order to use it, you will need to create the database from two files on the accompanying disk. Only the database structure is contained in the EXTRACT.DBF file on the disk that you just accessed. It contains the "skeleton" of the database without any of the data. Within dBASE, you can append data from a comma-delimited file into a preexisting database structure. After you have selected the EXTRACT.DBF database, you will need to exit from the Assistant or the Control Center by pressing **<esc>** until you have only the bottom status line and the dot prompt (".") visible. At the dot prompt type ***APPEND FROM EXTRACT.CDF TYPE DELIMITED*** and press **<enter>**. After you have appended the information into the database, the status line should show 408 records in the database .

3.9 DISPLAY/EDIT INFORMATION FROM THE DATABASE

Now we will look at the information in the database without the use of a screen format. If using the Assistant, select **Update** from the main menu and **Display** from the pull down menu; this will display the record where the pointer is. From the dot prompt, the ***Display*** command will perform the same operation. From the Control Center, pressing **<F2>** after selecting a database displays a range of data in tabular form. The tabular display is in the "browse" mode (see Fig. 3.7). In dBASE III Plus and from the dot prompt , the same type of display can be accessed by using **Browse** rather than **Display**. Both **Display** and **Browse** list the data to the screen in such a manner that it would be difficult to retrieve information from the database. A more pleasing screen can be obtained by using the **Edit** command rather than **Display**. Figure 3.8 is the default

Records   Organize   Fields   Go To   Exit

GROUP	SPECIMEN	REFNO	LATITUDE	LONGITUDE	ROCKNAME	SIO2	TIO2	AL
BTO	A	20	15.210	145.770	ANDESITE	57.40	0.62	18
BTN	I	20	-6.610	155.560	ANDESITE	62.70	0.63	16
BTN	H	20	-6.010	155.200	ANDESITE	62.20	0.59	16
BTN	G	20	-5.930	154.980	ANDESITE	60.80	0.77	16
BTN	F	20	-6.770	155.740	ANDESITE	58.80	0.81	16
BTN	E	20	-6.130	155.260	ANDESITE	56.50	0.40	17
BTN	D	20	-6.130	155.220	ANDESITE	56.30	0.17	17
BTN	C	20	-6.150	155.180	ANDESITE	55.20	0.90	17
BTN	B	20	-6.150	155.180	ANDESITE	54.80	0.92	17
BTN	A	20	-6.130	155.210	ANDESITE	54.00	0.27	16
BC	F	132	56.600	-133.830	ANDESITE	57.70	0.91	16
BR	D	149	22.820	74.000	BASALT	49.34	2.19	13
BR	A	149	23.200	69.900	BASALT	49.86	1.00	11
BY	A	153	60.200	25.530	GRANITE	73.78	0.24	11
BX	I	153	60.030	24.350	GRANITE	68.40	0.52	14
BX	H	153	60.050	24.020	GRANITE	74.76	0.02	13
BX	G	153	61.400	23.980	GRANITE	73.03	0.20	14

Browse ‖D:\dbiv\EXTRACT ‖Rec 1/408 ‖File ‖ ‖ Num

Figure 3.7 Browse mode of editing in dBASE IV. This is default mode from Control Center. In dBASE III Plus, same effect can be generated by BROWSE.

```
CURSOR    <-- -->       UP    DOWN      DELETE        Insert Mode:  Ins
 Char:     ←   →   Field:  ↑     ↓    Char:   Del     Exit/Save:    ^End
 Word:  Home End   Page:  PgUp  PgDn  Field:  ^Y      Abort:        Esc
                   Help:   F1         Record: ^U      Memo:         ^Home

GROUP        BTO
SPECIMEN     A
REFNO          20
LATITUDE      15.210
LONGITUDE     145.770
ROCKNAME     ANDESITE
SIO2         57.40
TIO2          0.62
AL2O3        18.40
FE2O3         4.88
FEO           2.26

EDIT          <D:> EXTRACT           Rec: 1/408          Num
```

Figure 3.8 Default EDIT screen under Assistant.

editing screen. Each of the fields is displayed in a vertical list and can be read easily. **Edit** allows you to edit any of the data displayed for that record and is the only form of seeing data on screen in a predetermined format. You also may edit in **Browse** which is especially advantageous when you wish to change the same item in a number of records.

Both **Browse** and **Display** portray the data in a fixed format that may be suitable for the user's purposes. **Edit** is the only command which allows the user to specify a screen format to be used. If no format is listed, the default single column format will be used. In order to use a screen format, the user must create one. The process for creating a screen format is similar to the process of modifying one.

The ability to alter the screen to conform to user preferences is needed both for data entry and retrieval. dBASE stores such information in separate files with the .FMT and .SCR extensions. The default form is a single vertical column of field names displayed on the left side of the screen and blocks of inverse video mark the space allocated for each to the right of the field name. By accessing the ***CREATE FORM*** command (**Format** under **Create** in the Assistant, **create** under **Forms** in the Control Center), the user can place the selected fields in any location on the screen and add additional information to the form. The processes are similar, but there are a few differences.

In performing this operation under dBASE IV, after selecting **create** under **Forms**, the user must first select a field (or fields) to be positioned on the form. This is done by selecting **Add field** under the **Fields** main menu item (see Fig. 3.9). A secondary screen will open and a list of fields will be displayed. In this example, select the ROCKNAME field (see Fig. 3.10). The appearance of this field can be altered before placing it in position or it can be moved in the "blackboard" mode (see Fig. 3.11). In this mode the screen shows where the field has been placed and allows it to be moved. The function keys that must be pressed to select, move, and copy fields are displayed on the bottom of the screen. After selecting the ROCKNAME field, pick it up and move it to the middle of the first row of the screen. In the blackboard mode this is done easily by moving the field to the location desired. The highlighted region will display only the contents of the field and not the field name itself. Notice that the field is present and will display the data in that position, but will not display the name of the field on the screen to know what information is being displayed in this block. Move the cursor to the side of the field and type ***Rock Name*** to identify it. In a similar manner, all of the other fields may be selected and positioned on the blackboard.

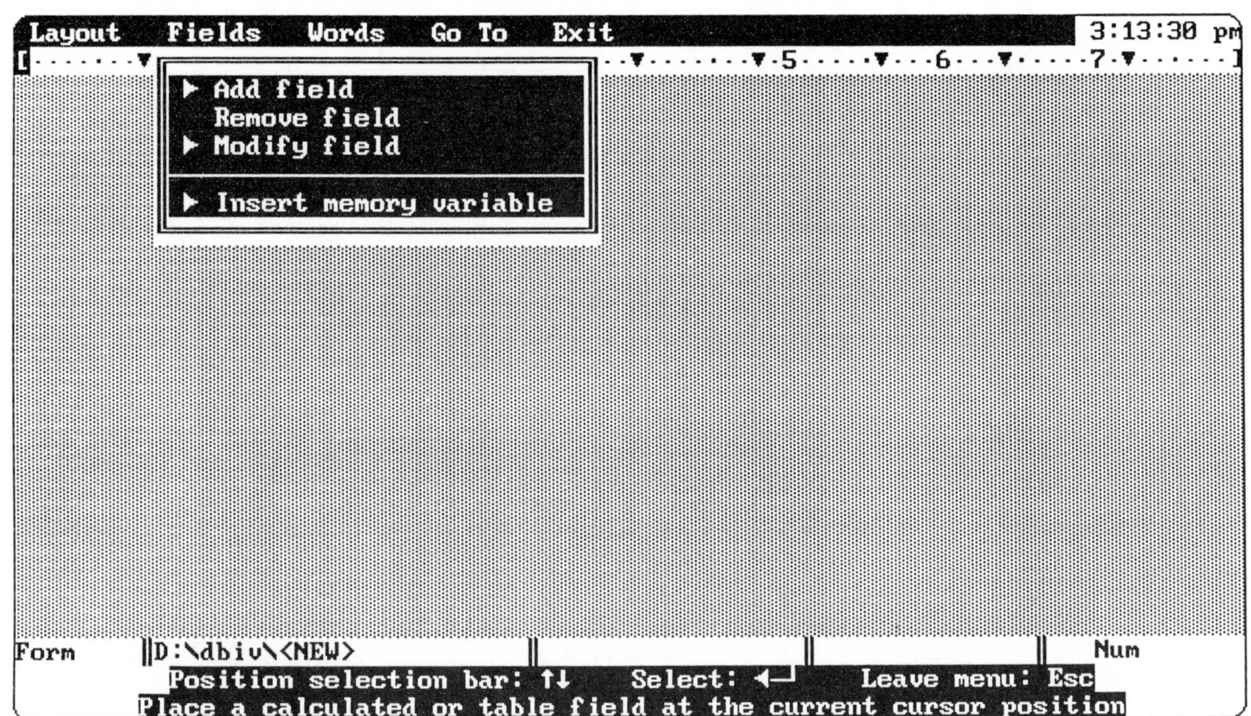

Figure 3.9 FORM screen under dBASE IV. After selecting <create> under Forms options, user then can enter fields by selecting Fields on this screen's menu.

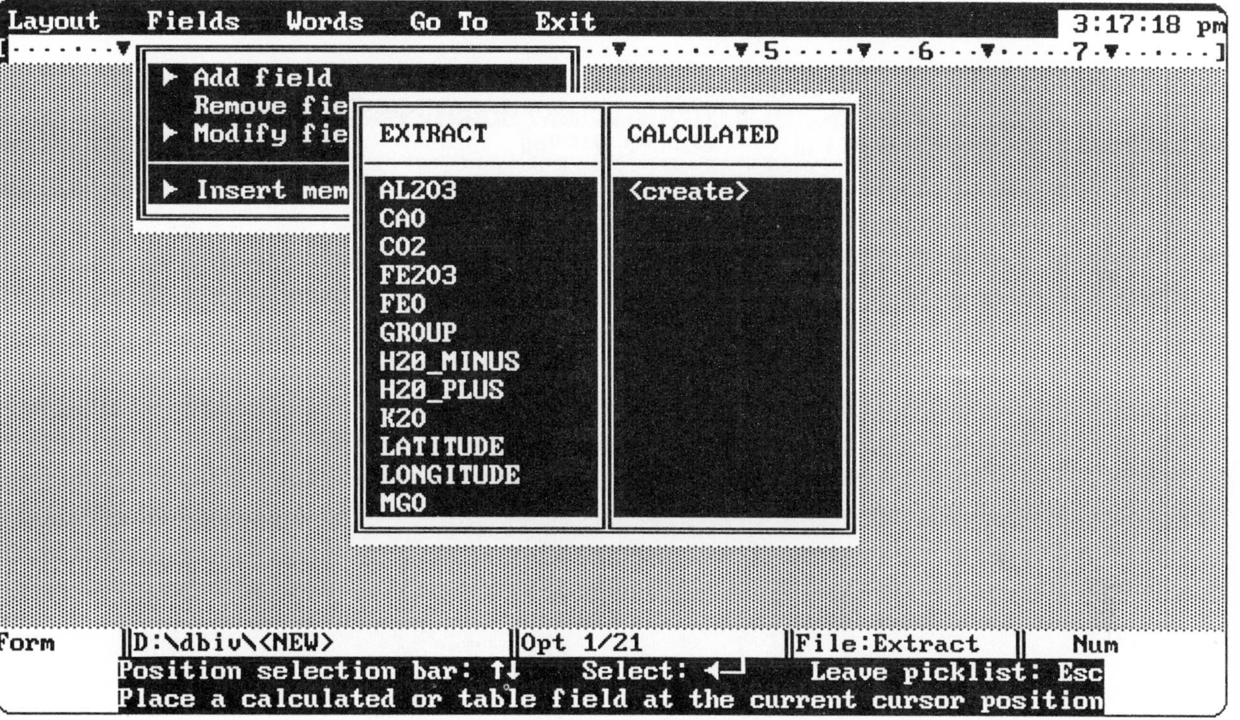

Figure 3.10 Selecting ROCKNAME field to add to screen form.

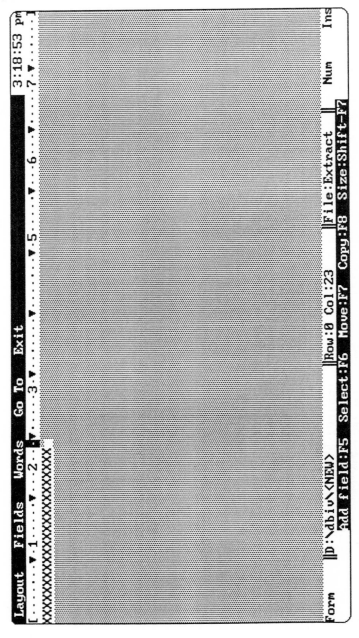

Figure 3.11 "Blackboard" mode of screen forms under dBASE IV.

Any additional text or highlighting lines also can be placed on the blackboard. Once the image of the form is correct, the entire format can be saved. Once saved, this format file can be used in displaying/editing the data. Any number of format files can be created and associated with a single database. However, only one format file can be invoked at any point in time. Under dBASE IV, you may use your stored format by selecting it from the **Forms** column. A secondary screen will ask you if you wish to **Display data** with this form or **Modify layout**. Altering a format file essentially is the same operation, only the original file will be open as a default format to be changed.

Under dBASE III Plus, you create a format file and save it. To do so, select **Format** under the **Create** menu item. After naming your format, the form layout screen will be displayed (see Fig. 3.12). Under **Set Up**, you should select what field(s) you wish to place on your form by selecting **Load Fields**. After you have selected the fields, switch to the blackboard mode via **<F10>**. As described above for dBASE IV, you can use the cursor controls to position the field to any location on the screen that you wish. The **Modify** menu item allows you to specify the format of the field to be displayed. After you have developed the format to your satisfaction, **Exit** and **Save** the results. In order to use the format, you must select **Format for Screen** under the **Set Up** menu item.

All of the operations described previously can be invoked from the dot prompt by typing ***CREATE SCREEN***. After giving dBASE the name of the a file to modify or a new name for a new format, you will see the same screen layout screens as before. In order to use the format, the dot prompt command is ***SET FORMAT TO ...*** Having set the format, try using it with the **Edit** command. Please note that with a format set, both **Browse** and **Display** remain unaffected.

Modifying a form is done in a similar manner. Please use the FORM1 file on your disk which will show the format as in Figure 3.13. In this situation, modify the form to have two columns of oxides rather than three. In order to do so, we should copy all of the FORM1 files under a new name so as to preserve the originals. Please be sure to copy and rename all of the files with the FORM1 prefix. Having done so, please add this file to your catalog, if you are using dBASE IV. The new format file should be displayed on the list of current formats to be selected under the Assistant or from within the Control Center. Select your copy and select **Modify layout** under the Control Center, or **Format** under the **Modify** menu item in the Assistant. The dot prompt command is ***MODIFY SCREEN <filename>.*** You should now be at the same screen as form

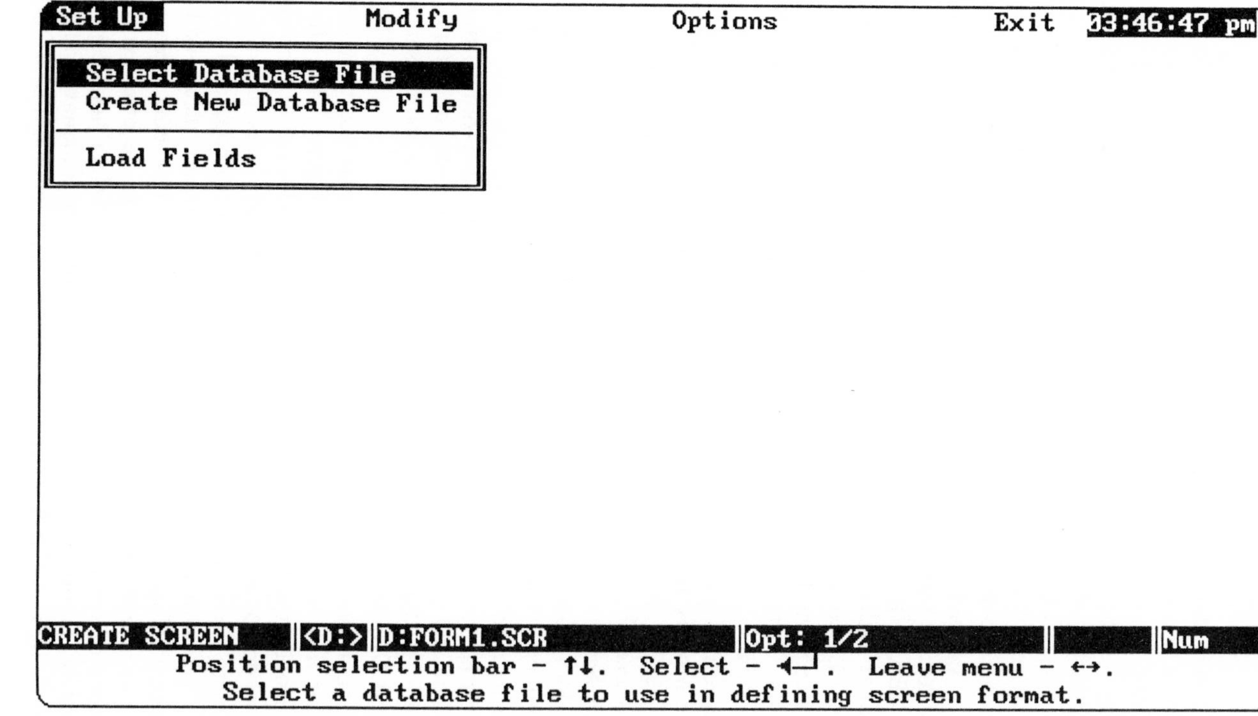

Figure 3.12 FORM screen under Assistant. This screen is analogous to Figure 3.9 under dBASE IV.

```
Set Up             Modify                Options              Exit  04:03:03 pm
                Group  XXXX      Specimen  XX    Reference No.  9999

                   Latitude  9999.999    Longitude  9999.999

                      Rock Name    XXXXXXXXXXXXXXXXXXXXXX

                Si02      99.99       Mn0      99.99    P205       99.99
                Ti02      99.99       Mg0      99.99    C02        99.99
                Al203     99.99       Ca0      99.99    H20_PLUS   99.99
                Fe203     99.99       Na20     99.99    H20_MINUS  99.99
                Fe0       99.99       K20      99.99



              Status Symbols:  XXXXXXXXXXXXXXXXXXXXXXXXXXXXXXXXXXXXXXX

CREATE SCREEN  ║<D:>║D:FORM1.SCR          ║Pg 01 Row 17 Col 11║Ins   ║Num
      Enter text.  Drag field or box under cursor with ↵.  F10 for menu.
                       Screen field definition blackboard
```

Figure 3.13 FORM1 screen.

creation, only the blackboard has the old format on it already. Please select the STATUS SYMBOL field and move it down four lines. Then select each of the oxide fields sequentially and move them into a two-column arrangement. When finished, the form should resemble Figure 3.14 and you should be able to create/modify any screen format for your own use.

3.10 MODIFYING OR CREATING A DATABASE

The two processes of modifying and creating any kind of file, including a database, essentially are the same under different versions of dBASE. The primary operation under **Modify database file** or **Create Database file** is to select what information will be contained within the database. After creating a database, fields can be added or removed. However, if a database has information already within its structure, adding a new field will add a blank entry for each record in the database. This implies that the users will enter all of the additional data manually for the preexisting records. If you are creating your own database and do not have a disk space problem, include all possible fields of interest. As you gain experience and as your database(s) become larger, the decision about what to include and where to include it becomes more cluttered by the existence of other options. . As a novice, however, a good rule of thumb is "if any item of information may be useful, include it". When linking databases into a relational form, this advice will need to be revised.

For each field entered, you must name the field (maximum ten characters with no blanks, starting with a letter), define the field type, and fill in some description about the field characteristics such as number of decimal places, etc. The order in which you enter the fields determines how data will be entered into that form, if done manually. It is a good idea to design a form for data entry on paper before entering the fields and their definitions into dBASE. Once you have decided in what order the data will be entered on the form, the order in which the fields should be entered when creating the database will be apparent. Although it is not difficult to alter the list later, it may prove bothersome to maintain data already entered or you may create an entry form that "skips" to inappropriate fields. Planning prior to creating a database forces the programmer to also make decisions about what should be included in the

```
Set Up              Modify                    Options                  Exit  04:14:44 pm
               Group  XXXX       Specimen  XX    Reference No.  9999

                     Latitude  9999.999    Longitude  9999.999

                       Rock Name   XXXXXXXXXXXXXXXXXXXXXXX

                     SiO2      99.99              CaO           99.99
                     TiO2      99.99              Na2O          99.99
                     Al2O3     99.99              K2O           99.99
                     Fe2O3     99.99              P2O5          99.99
                     FeO       99.99              CO2           99.99
                     MnO       99.99              H2O_PLUS      99.99
                     MgO       99.99              H2O_MINUS     99.99

              Status Symbols:  XXXXXXXXXXXXXXXXXXXXXXXXXXXXXXXXXXXXXXXX

MODIFY SCREEN   ||<D:>||D:FORM2.SCR          ||Pg 01 Row 13 Col 21||Ins   ||Num
    Enter text.  Drag field or box under cursor with ◄┘.  F10 for menu.
                       Screen field definition blackboard
```

Figure 3.14 FORM2 screen modification of FORM1.

database, what the range of appropriate or possible values will be, and what is the "normal" order of entry of such data. The greater the amount of thought about the structure and nature of the database before creating it, the better the results will be. The structure can then be saved and data can then be entered by hand or imported from a machine readable file.

As an exercise, let us create a new database that only contains the rockname and five major oxide contents of the specimens in the EXTRACT database. Using the Control Center, select **create** under the **Data** item. The first field to enter will be the name of the rock. This information will be character in nature and because we are following the EXTRACT database, it will have a maximum number of 23 characters. As you type in **ROCKNAME,** dBASE IV will assume that this field is a character field and is not an index field. By pressing **<enter>**, you will accept these conditions as a default and eventually enter the width of the field. When on this line, you can use the cursor arrows to move around and make changes. Once you press **<enter>** in the last entry (**Index**), you will move on to the next field. By using the up arrow, you can return to edit a prior first if need be. The next field will be a numeric field to contain data on the SiO_2 weight percent. Any name can be used, SILICA, SI, SIO2, etc. However, the numerical value that will be in this field, if taken from EXTRACT, is the SiO_2 weight percent. All of the other names could be ambiguous. The decision is yours, but remember that if anyone else uses your database, they will need to know what type of data this field contains. Having named the field, press **<enter>** to move to the **Field Type** position. You then can scroll the choices using the space bar. You will notice that each type of data field is described at the bottom of the screen. The descriptions of Float and Numeric fields are exactly the same. However, dBASE IV does not treat them exactly in the same manner. Float values are stored as floating point numbers and allow you to characterize better large and small numeric values. Numeric values are stored as binary coded values with a much smaller range of possible values. The gain in using Numeric fields is that greater precision can be obtained. Since these values are all between 0 and 100, we will use Numeric fields for them. After selecting the Numeric field type, we must also enter a field width and decimal location. As per the instructions on the bottom of the page, we should count the decimal point and sign as characters in the field width. The data set has weight percents to the nearest hundredth, so the maximum value possible is 100.00. Since all of the values are positive, we do not have to allow for the sign location. Therefore the width to be entered is 6 and the decimal places is 2.

Repeat this operation to create TiO2, Al2O3, Fe2O3 and FeO fields in your new database. When you have completed the structure, select **Save changes**, exit using the **Exit** menu item, and give your new database a name. If your are unsure of your result, the SMALLEXT database structure on your disk is an example of what you should have in your database.

Creating a database in dBASE III Plus is a similar operation to that of the Control Center in dBASE IV. Under the Assistant, you would select **Database** under the **Create** menu item. The program will inquire where the new database is to reside and its name. At the top of the screen you will see instructions on what keystrokes will perform specified operations. As stated previously, type in **ROCKNAME** for the first field, press **<enter>** and accept **Character** as the type by pressing **<enter>**. Enter a width of 23 and by pressing **<enter>** you will complete the first field and start the second. Enter **SIO2** and press **<enter>** for the second field name. Use the space bar to move the **Type** to **Numeric**, then press **<enter>** to accept it. As before, the width should be 6 and the decimal places should be 2. Enter the other fields for the major oxides in a similar manner. When the structure is complete press **<Ctrl><End>** simultaneously. This is marked as **Exit/Save** in the help area at the top of the screen. dBASE III Plus will ask for confirmation **(Y)** and then ask if you wish to input data records now **(N)**.

3.11 MANEUVERING WITHIN A DATABASE

Having created a database, you can enter data manually or transfer data from a machine readable file (as we did to create the EXTRACT database) into your new database. With data now stored in the structure that you created, the next step is to maneuver manually around the database and alter the values in the respective fields as need be. Let us use the EXTRACT database that we created originally. Make this database the active file by selecting it under **Data** in the Control Center or by **Database File** under **Set Up** in the Assistant. From within the Control Center the easier method of editing and maneuvering within the database is via the **Data** option **<F2>**. This places you into an editing mode (see Fig. 3.15). Items in any field can be altered by typing over the current entry. **<Pg Dn>** can be used to move from one screen to the next

```
 Records   Organize   Go To   Exit
GROUP        BTO
SPECIMEN     A
REFNO          20
LATITUDE      15.210
LONGITUDE     145.770
ROCKNAME     ANDESITE
SIO2         57.40
TIO2          0.62
AL2O3        18.40
FE2O3         4.88
FEO           2.26
MNO           0.00
MGO           2.40
CAO           8.48
NA2O          2.76
K2O           0.74
P2O5          0.00
CO2           0.00
H2O_PLUS      1.30
H2O_MINUS     0.00
STATUS_SYM   2A,3C,3I
Edit    ║D:\dbiv\EXTRACT          ║Rec 1/408          ║File ║          ║   Num
```

Figure 3.15 Default data editing mode under Control Center.

and to additional records within the database. The **Records** item in the main menu also has an option for **Undo change to record**. In order to move the record pointer quickly, you need to select an item under the **Go To** main menu item. The four items on the top of the supplemental screen are for simple maneuvers. The **Top** and **Last** records are self-explanatory. **Record Number** allows the user to go to a record at a particular position in the database, that is go to record No. 23. The **Skip** command is used to browse through the database by jumping a specific number of records. The bottom four items are for having dBASE IV search the database to find a record that matches a particular condition. Once this condition is met, dBASE IV will leave you at that record to edit it. Using these conditions, let us locate the first occurrence of basalt. Move the cursor to the ROCKNAME field under the first entry (ANDESITE) then access the **Go To** menu item and the **Forward Search** item in the pop-up menu. When the program asks for a search string, enter ***BASALT***. Please be sure that you enter BASALT all in uppercase letters. You can alter this feature by selecting the last line in the popup menu box, **Match capitalization**. If this entry reads **No**, then any match will work. The default however is set to yes, so if you do not change it, BASALT will match, basalt will not. If you do not match the entry perfectly, you will not jump to a new record. If all goes well, you should be on record No.12. Notice that after you have made a search, the criteria will remain in the **Forward** and **Backward** search criteria for you to locate the next (or previous) record that matches.

Moving around in the database with the Assistant in dBASE III Plus can be accomplished quickly by using **Position** on the main menu. Notice that after making EXTRACT the active database, dBASE III Plus will only allow you to **Locate**, **Skip** or **Goto Record**. **Seek** is used only in indexed files. **Goto Record** and **Skip** work as described previously. **Locate** determines the next record that meets a set of conditions that you can specify. Specifying conditions within the Assistant is the same for many operations. After selecting **Locate**, a pop-up menu will appear with five entries(see Fig. 3.16). As you select any entry you will be building a ***LOCATE*** command above the status line at the bottom of the screen. To perform the same operation as described earlier, select **Build a search condition.** Once this selection is made secondary areas appear with a list of field names and a description of the highlighted field (see Fig. 3.17). Using the cursor arrows, position the selection bar on **ROCKNAME** and press **<enter>** to select it. A new pop-up menu with search conditions will be displayed. In this example, we will be using the **=Equal To** option.

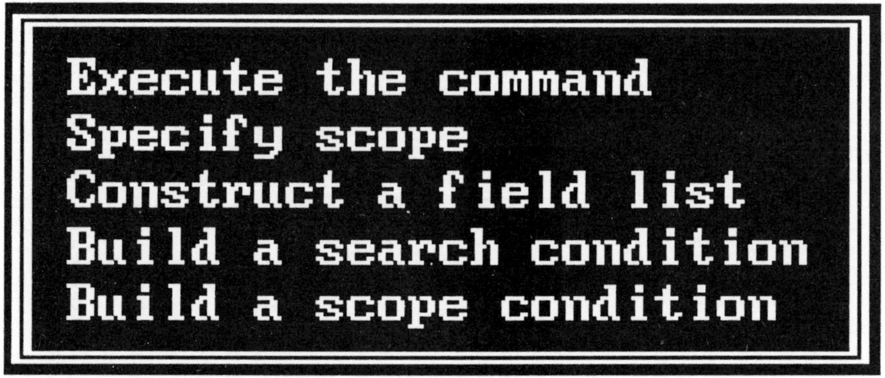

```
Execute the command
Specify scope
Construct a field list
Build a search condition
Build a scope condition
```

Figure 3.16 Secondary menu to Locate command under Assistant. This secondary menu of five items is displayed whenever a search is selected. Items within secondary menu may not all be available depending upon type of search requested.

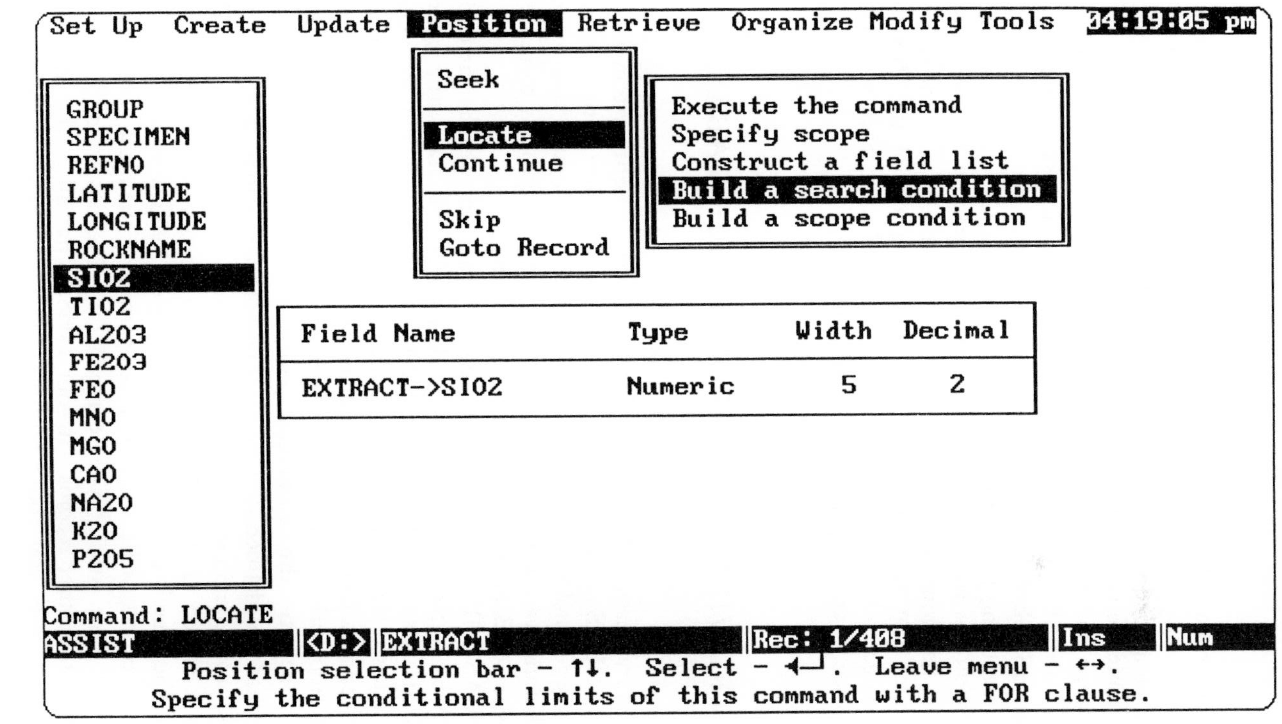

Figure 3.17 Building search condition for SIO2 in dBASE III Plus.

The Assistant then will ask for a comparison string to work with, type in ***BASALT.*** As before the use of all capital letters is important. Since we are using a simple condition, no other clauses will be used, so accept the **No more conditions** option to return to the first secondary menu. The **Specify scope** and **Build a scope condition** can be used to limit the records that dBASE will test the search condition against. For example, you can tell dBASE to test only the search condition against the next 30 records by specifying the scope to be that set of records. In this example, no other conditions will be used, so you may now **Execute the Command**. If everything has gone well, at the bottom of the screen you will see that the first match was at record No. 12. After using any of the **Position** commands to go to a particular record, you will need to go back to **Update** and **Edit** to see what that particular record contains.

3.12 RETRIEVING DATA FROM THE DATABASE

The primary reason for using a database management system is the ability of a DBMS to retrieve particular pieces of information from a large data set quickly and efficiently. After the simple maneuvers described before, we are ready to begin automating the process to retrieve different types of data from the EXTRACT database.

With dBASE III Plus there are several ways in which particular records can be retrieved automatically. Using the Assistant, the easiest way is to **List** those records that meet certain criteria. When you select **List** , you will see the same screen as with Locate (see Fig. 3.16). Each of the five selections, **Execute the command, Specify the scope, Construct a field list**, and **Build a scope condition**, performs different operations. Of course, **Execute the command** performs the operation under the conditions that you defined. There are two commands that determine the scope of the search and two commands that specify search conditions. The scope of the search is defined as that group of possible records to be checked for the search conditions. For example, if you wish to search the database for records of basalts that meet particular chemical composition tests, it would save time if you searched only the basalts and did not bother testing the other records. If your database has all of the basalts at the end of the base, you could position the record pointer to the first basalt record and then specify the scope as being "REST". You can specify a particular record or the most generally used scope, "ALL" records. If the basalts were not all in one section, you could specify a different scope condition exactly the same way as setting a search condition. A scope condition allows you to restrict the search conditions to only those records that first meet the scope condition criteria. Continuing our example, you could achieve the same results by setting the scope condition to ROCKNAME="BASALT". The program would first select records where that condition existed, then test for the search condition. The test conditions presented in the Assistant are equivalence (=), greater(or less) than(> or <), greater (or less) than or equivalent (>= or <=), and not equivalent to (<>). In each situation, the user specifies a field in the database, then the test conditions are selected from the menu. For example, first you might select **SIO2** as the field, then "<=" from the next menu, followed by "46" to set a condition of "SiO2<=46". If the condition is met, the record is included in the scope, or in the output if this were a search condition.

Compound and complex conditions can be created from more than one test condition. Each condition has a logical value of either true or false. If the record satisfies the condition, that is the SiO_2 content is 45 for the previous statement, then the condition has a value of True. Otherwise the condition has a value of False. Records are selected if the overall value of the statement is true based upon the values of the conditions that comprise that statement. If only one condition is used and that condition is True, that particular record will be selected. A logical relationship between multiple test conditions is specified in a compound conditional statement. This entails using .AND. or .OR. as selected from the menu. If an .AND. is used, both clauses are linked by the .AND. must be True to return an overall value of True. For example, if you used "SiO2<=46 .AND. Al2O3>12", a record passing the first condition would give that clause a value of True. If the record passed the second part of the test, "Al2O3>12", then that clause also would have a value of True. This makes the entire statement True. If either (or both) of the two clauses were false, however, the combined condition would be False. The conjunction .OR. works in the different manner, if either of the two clauses is true, then the overall statement is True. For example, if you used "SiO2<=46 .OR. Al2O3>12" and a record met either of these conditions, the combination would be True. Each of the command line statements built by this process can be invoked from the dot prompt of either dBASE III Plus or dBASE IV. In dBASE III Plus, queries can also be stored in a file format so that a query can be repeated easily. This is analogous to the method used in the Control Center to locate a group of records.

From within the Control Center of dBASE IV, you can create a query file to act as a filter to select a subset of records from the database that meet a given set of conditions. This is performed by a query-by-example (QBE) system built into dBASE IV. Make sure that you have selected a database to use, then select **create** under the **Queries** item. The QBE screen will have the database and fields loaded as shown in Figure 3.18. The presence of a down arrow shows that a field has been selected to be part of the filter to be used. Use the **<Tab>** to choose a field and press **<F5>** to include it in the query. The conditions that must be met are written in the column below the selected item. For example, to select the basalt records, the rockname field must be selected and ***BASALT*** entered in the column below. By pressing **<F2>**, the query is invoked and the results displayed. Please note that only the fields selected will be visible and that the display will be a **Browse** mode of editing. When two or more conditions are placed upon the same line, an AND connection is assumed. By entering <50 under SIO2 and pressing **<F2>**, the search will be based upon samples that are basalts and have SiO2 values < 50%. The OR situation is handled by placing the entries on different lines in the query form.

3.13 SEARCH EXERCISES

Let us perform a series of searches on the EXTRACT database as examples. Using the Assistant, we wish to retrieve analyses that have SiO_2 values > 70%. We need only build a search condition as we have done before. When you have selected this option and identified SIO2 as the field chosen, the screen should be displayed as in Figure 3.19. The relationship menu then will be displayed under the search condition menu. Using the cursor, select greater than (>) from the menu of choices and press **<enter>**. For the numeric value, type *70*. Remember that the dot prompt language command line is being built just above the status line as you select each part of this command. At this point we are not using any other conditions, so select **No more conditions** and you will be returned to the search conditions menu. Since there is nothing else that we wish to use, **Execute the command** at this time. Above the status line you

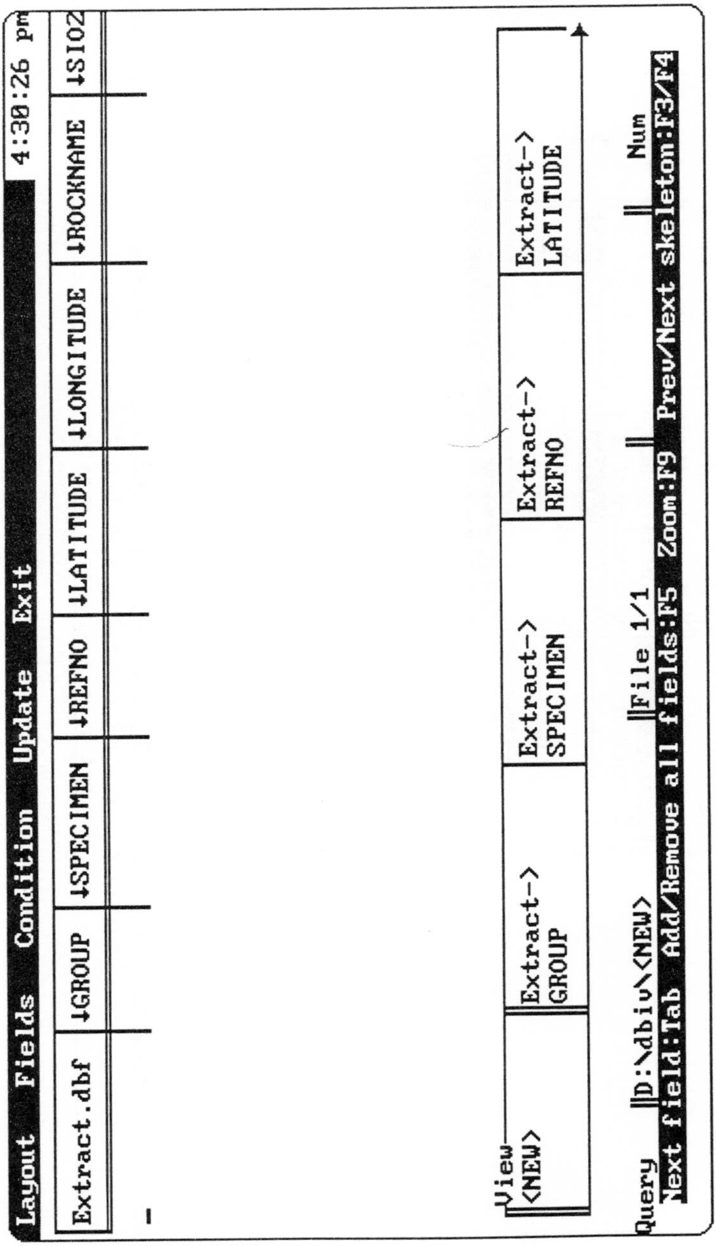

Figure 3.18 Query-by-example (QBE) screen from dBASE IV.

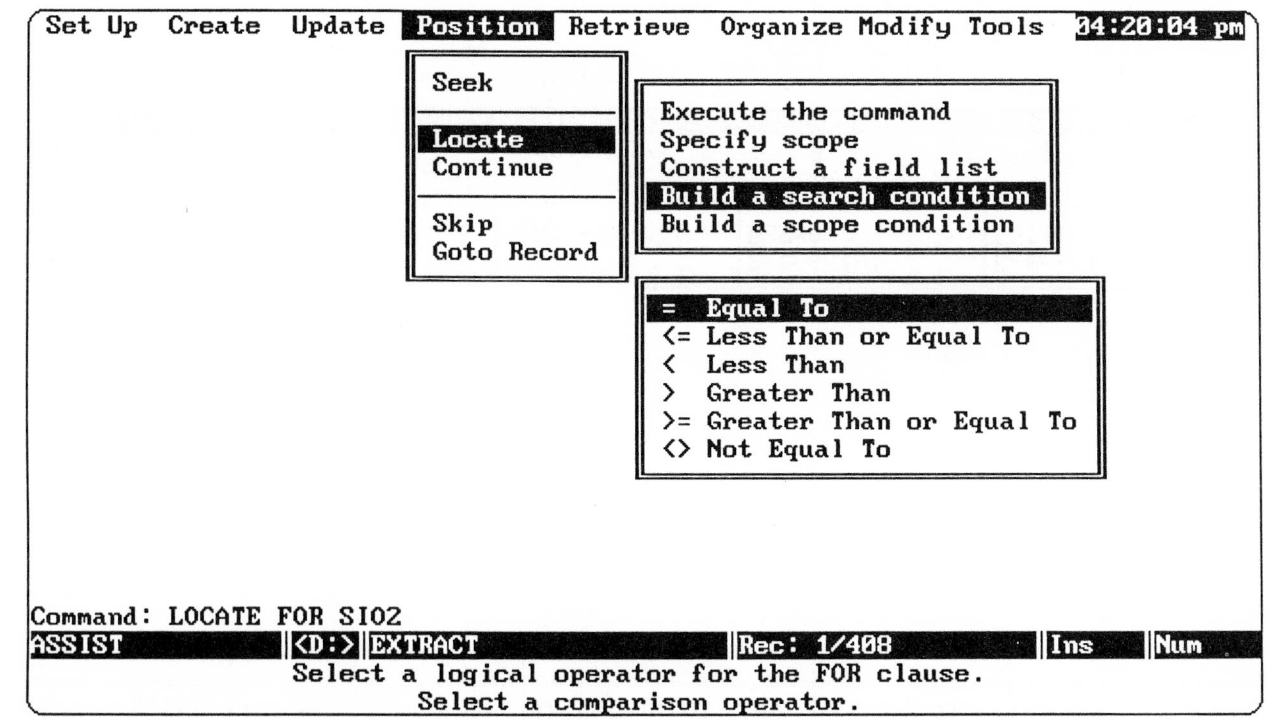

Figure 3.19 Search for SIO2>70 as command is built.

will see that dBASE has determined that record No. 14 meets our requirements of > 70% silica. dBASE also leaves us at that record position, so that we can move over to **Edit** to see our record. To locate the next record, move back to **Position** and use **Continue** to continue to locate the next record that meets the criteria. Once you have done this tedious technique, select Retrieve from the main menu and either **List** or **Display** from the pull down menu. Please repeat the same search conditions for SiO2>70%. **Sum, Average,** and **Count** also can be used to retrieve information. You can "count" the number of records that meet the SiO_2>70 criteria (74 records). To find the average Al_2O_3 from those records, specify the search condition as before and use **Construct a Field List** to specify Al2O3. If you do not specify a field list, then you will get the average of each field. Notice that the answer shows you how many records were averaged(74 records) and that the answer is 13.06. Search the EXTRACT database for the following information:

(1) How many samples have SiO_2>50% and what is their average Al_2O_3 content? (Answer: 289 records with an average Al2O3 of 15.17).

(2) How many samples have $SiO_2 \leq 65\%$? ... and which records meet this criteria? (Answer: 307 records using **List** would show each record).

(3) How many samples have SiO_2 values between 50 and 65% inclusive? (Answer: 191 using SIO2>=50.AND.SIO2<=65).

(4) What is the average SiO_2 content of the basalts in this database? (Answer: 49.54 for 188 basalts using **AVERAGE FOR ROCKNAME = 'BASALT' SIO2**).

Using the Control Center in dBASE IV, one can perform the same operations using the QBE system. After selecting the EXTRACT database, select **Create** under the **Queries** heading. All of the fields of the EXTRACT database will be included in the resultant screens. If you wish to modify this, please use the **<Tab>** to move to the field that you wish to remove and use **<F5>** to remove(or add) the field to the view. Move the cursor to the SIO2 field and enter **> 70** on the first line below it. Pressing **<F2>** will process the query and show you all of the records that meet your criteria. The screen should resemble Figure 3.20.

Records Organize Fields Go To Exit

GROUP	SPECIMEN	REFNO	LATITUDE	LONGITUDE	ROCKNAME	SIO2	TIO2	AL
BY	A	153	60.200	25.530	GRANITE	73.78	0.24	11
BX	H	153	60.050	24.020	GRANITE	74.76	0.02	13
BX	G	153	61.400	23.980	GRANITE	73.03	0.20	14
BX	E	153	60.220	23.550	GRANITE	71.69	0.34	13
BX	B	153	60.470	21.080	GRANITE	72.63	0.29	13
QB	D	154	44.950	-72.530	GRANITE	73.41	0.19	14
KH	F	404	-24.800	29.800	GRANITE	71.80	0.39	12
KH	E	404	-25.200	27.900	GRANITE	74.06	0.24	12
NW	F	514	44.500	-77.600	GRANITE	75.88	0.19	11
NW	E	514	44.500	-77.600	GRANITE	75.45	0.23	11
BSG	N	1762	58.000	-4.000	GRANITE	73.68	0.20	13
BSG	G	1762	58.000	-4.000	GRANITE	74.89	0.00	15
BSG	F	1762	58.000	-4.000	GRANITE	74.17	0.06	15
BSG	E	1762	58.000	-4.000	GRANITE	72.85	0.32	15
BSG	D	1762	58.000	-4.000	GRANITE	74.41	0.15	14
BSG	C	1762	58.000	-4.000	GRANITE	72.55	0.06	15
BSG	A	1762	58.000	-4.000	GRANITE	71.77	0.05	16

Browse ‖D:\dbiv\<NEW> ‖Rec 14/408 ‖View ‖ ‖ Num

Figure 3.20 Results of query under dBASE IV.

A query file also can be created under the Assistant, by selecting **Query** under the **Create** main menu item. The same types of questions must be answered under the Assistant as from within the Control Center. The user selects a field and a condition from a series of screens that are similar to the secondary screens for **List** and **Locate**. The dot prompt versions of **List** and **Locate** as well as the query can be used in either form of dBASE. The single difference is that with a query, the filter components are stored as a file for repeated use.

3.14 ORGANIZING THE DATABASE

The use of a microcomputer-based database management system allows the user to keep large amounts of data in a readily accessible form. However, as the size of the database grows, the time it takes the computer to search the data sequentially also will increase. For the same reasons that good organization is an asset in locating information by nonelectronic means, organizing the database becomes more important with larger quantities of data. The degree of organization depends upon the usage of the database and the foresight of the user. The two primary techniques for organizing the database are **Sort** and **Index.**

Sorting the database is the first level of organization. This process mimics what a person with any large data set would do. At first, the user merely collects data. As the data set becomes large, the user will organize it in some manner, using sample numbers or some other key-based system. At this point, the data file can be sorted so that the records are kept in ascending (or descending) sample number order. In a database management system the process is the same. The user selects a database to be sorted and selects the field or field(s) to base the sort upon. The sorted database is a copy of the original file except that the positions of the records have been changed so that they now are in sorted order. Because it is a copy, most systems allow the user to store the sorted version under a name different than the original. The user then can decide to use the original version or the sorted version of the database.

The sorted version of the database can be searched faster, if the primary field of the sort is the same as the field in the search or scope condition. For example, let us assume that you will be accessing the EXTRACT database using SIO2 as a common field in your searches.

First, we would have to copy the database into a sorted form. In the Assistant, this is done by selecting **Sort** in the **Organize** main menu item. Having selected **Sort,** the Assistant will ask you which field(s) to use. You could specify a series of fields to use to correctly position any records that have the same primary field value. In this situation, let us just select **SIO2.** We then have to name the sorted version of the database. Please be sure not to name it EXTRACT and please add the .DBF extension, so that dBASE will see it as a database file in its own right. Since dBASE must sort the database and write out a copy, sorting does take some time depending upon the database size. To view the result, you will need to go to **Set Up** to get the new database and **Edit** to see each of the records. Please note the SIO2 values as you scan the records. The first record should be 40.25, then 40.26 and so on.

Under the Control Center in dBASE IV, this operation can be performed after a database is selected. The second screen after a user selects a database allows a selection, **Modify the structure/order.** Sorting the database is altering the record order to create a new database. The initial screen has a main menu item **Organize** and under it is **Sort database on field list.** By accessing this combination you may select **SIO2** and **Ascending order** to create a sorted version of the EXTRACT database.

Although a sorted database can improve the search times of a DBMS, an indexed file can improve performance even more. The index operation creates a separate file associated with the database that contains the positions in the database of records sequenced by a user selected specific criteria. In **Sort**, the user copies the entire record physically into a new location thereby doubling the space on his disk that contains the information. Under **Index** records remain in their original locations, but their positions based upon the index key field are recorded. The differences can be seen in Figure 3.21. Multiple indices of a database can be created. The size of an index file is small compared to the entire database. Use of an index speeds up the processes of database operations because the index can be scanned quickly to find records satisfying numerous criteria.

Let us create an index based upon SiO_2 for the EXTRACT database. In the Assistant, you need only select **Index** under the **Organize** main menu item. It will then ask you for an index expression. In our simple example, select **SIO2** from the field list. You then will need to name the index file (extension of .NDX). The index is invoked when the database is selected under **Set Up**. Specify that EXTRACT is indexed and select

Original

Record Pos.	Group	Spec_ID	Ref.#	Rockname	SiO_2
1	BYY	D	425	PICRITE BASALT	46.60
2	NN	B	50	RHYOLITE	67.79
3	BYJ	G	392	GLASS	73.75
4	BYJ	A	392	BASALT	49.88
5	NN	A	50	RHYOLITE	75.11
6	BYJ	B	392	BASALT	47.63
7	BYJ	D	392	BASALT	46.93
8	BYY	L	425	BASALT	45.56
9	NN	C	50	RHYODACITE	65.58
10	BYY	G	425	3-PHENO. BASALT	46.80

A.

Record Pos.	Group	Spec_ID	Ref.#	Rockname	SiO_2
1	NN	A	50	RHYOLITE	75.11
2	NN	B	50	RHYOLITE	67.79
3	NN	C	50	RHYODACITE	65.58
4	BYJ	A	392	BASALT	49.88
5	BYJ	B	392	BASALT	47.63
6	BYJ	D	392	BASALT	46.93
7	BYJ	G	392	GLASS	73.75
8	BYY	D	425	PICRITE BASALT	46.60
9	BYY	G	425	3-PHENO. BASALT	46.80
10	BYY	L	425	BASALT	45.56

B.

SiO_2	Record Position
45.56	10
46.60	8
46.80	9
46.93	3
47.63	2
49.88	1
65.58	7
67.79	6
73.75	4
75.11	5

Figure 3.21 Indexed(A) vs sorted(B) databases.

the index file that you have just created. Please note that the program now is looking at record No. 331 of 408, which has the lowest SIO2 value in the database. As you go to the next record, the counter actually will go backwards to record 330, which has the second lowest SIO2 value. Although the database has not changed, you are looking at the information in a new sequence determined by your index.

dBASE IV has other indexing features that dBASE III Plus lacks. Simple indices can be created under the Control Center in dBASE IV in a process similar to that of creating a sorted version of the database. dBASE IV however allows automatic indexing of a database when you create it by marking fields to be indexed. dBASE IV also has a Master Index capability which allows the user to collect up to 43 indices together under one file heading (Lima, 1989).

Although the use of **Sort** and **Index** can improve the performance of your system, their incorrect usage also can increase the search time as well. Although a sorted database may improve searching based upon the primary field of interest, searching on other fields may become more time-consuming. Sequencing records by the value of one field may not improve the rate at which searches based on other fields can be performed. The use of indices carries the same type of problem. However, creating multiple indices and storing them for particular uses is less space consuming then storing multiple copies of the entire database. Both types of organization work well when you have some knowledge about the data set that you are working with and what exactly you want to do with it. However, another reason for indexing databases is the ability to link multiple databases.

3.15 CREATING LINKAGES BETWEEN DATABASES

A view is a structure by which several databases can be linked logically together. Views can be created in dBASE III Plus and dBASE IV. For example, EXTRACT, also has information about the literature reference from which the data is taken. The REFNO field has a number that corresponds to an entry in the REF.CDF file that has all of the bibliographic information. We could have stored the reference data in the original EXTRACT database. Each sample would have had author, source, date, etc. Unfortunately, this structure also would waste a great deal of space, because each reference actually has several samples in the EXTRACT database. Each sample from a particular reference would cause us to repeat that information in the database. By keeping it outside the EXTRACT structure, we will need to provide a linkage that will allow us to locate the reference after we have located a particular sample.

The simplest way to do this is to create an index based upon the characteristic(s) that link the two databases. First, we need to create a reference database. As before, use the REF.DBF structure and the **Append** command to fill the structure with the REF.CDF file. After the database is created, look at the structure and some of its records to find the linking field to the EXTRACT database, that is the reference numbers. Having determined that NUMBER in the REF database is the reference number, we then need to create an index of REF based upon NUMBER. This index file then will contain the positions of each entry in the REF database based upon its reference number. Once we have located a record in the EXTRACT database, we can use the REFNO entry in it and the index file for REF we just created to find the reference from which data has been taken. This linkage could be performed manually, but both dBASEs allow you to automate this process by being able to create and save a view.

3.16 CREATING A VIEW

dBASE is not a true relational database management system, but it does allow linking databases through the use of a view file. The user creates a view file which relates multiple databases through user-specified fields contained in each. The structure however is not open ended. Files may be linked in a sequential or linear mode, but not in a ring or circular structure (see Fig. 3.22). In most situations in geology, this constraint is not a drawback. If you are creating databases that will be linked at a later date, it is helpful to formulate the linkage pattern early in the development of the databases.

The creation of a view file is straightforward. The first operation is to select the databases to be linked and create the requisite indices as described earlier. In order to see data from multiple databases, a format for a composite screen also must be created. We will now create a view file for the EXTRACT and REF databases. Each entry in the EXTRACT

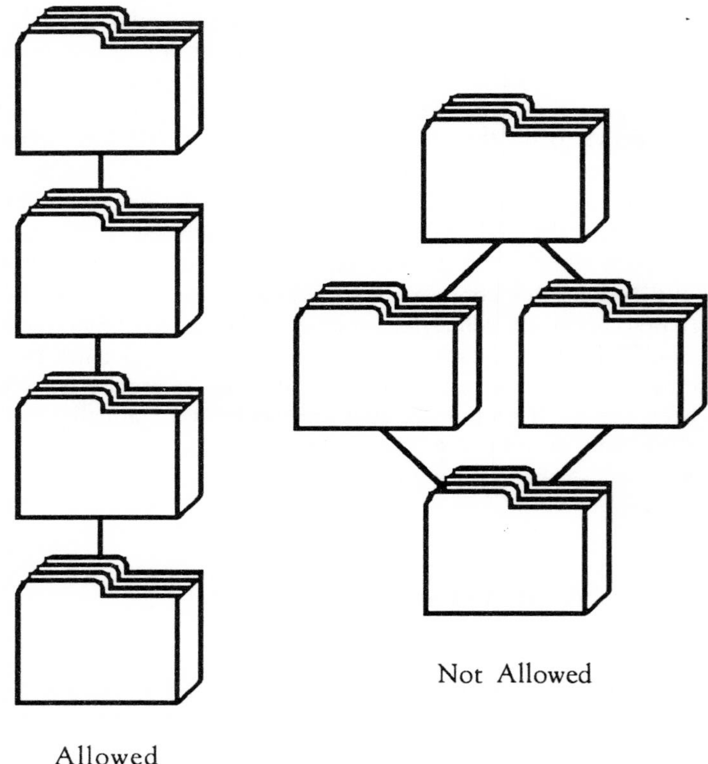

Figure 3.22 Database relationships allowed under dBASE. Linear chains and branching are allowed. Circular references are not allowed.

database will have one and only one entry in the REF database. However, many entries in the EXTRACT database will be linked to the same entry in the REF database. This sometimes is called a many-to-one relationship. The structure that we will now begin to create is shown in Figure 3.23. The

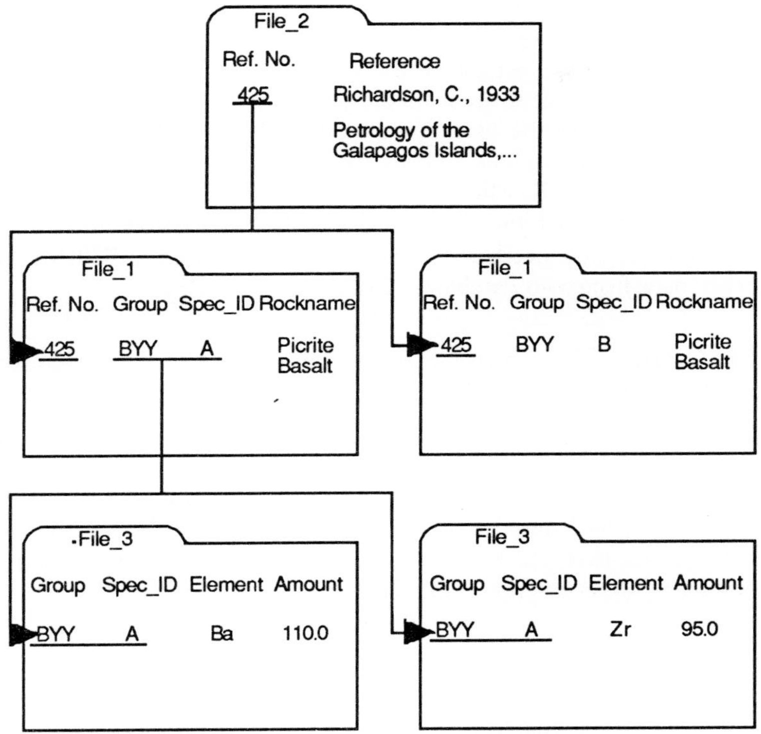

Figure 3.23 Linkages between databases. In example given you will be linking REF (File_2) database with EXTRACT (File_1) database. Trace element database (File_3) also could be linked into this structure.

steps to perform this operation under the Assistant in dBASE III Plus are:

(1) All of the databases to be included in the view must be indexed. The REF database should be indexed by NUMBER to provide the necessary linkage to EXTRACT'S REFNO field. EXTRACT can be

(2) A composite form with data from EXTRACT and from REF must be created. Copy the FORM1 files created earlier as FORM2.

(3) Select **Format** under the **Modify** main menu item and select **FORM2**. Remove some of the blank lines to make enough space for the reference data at the bottom of the screen. Use **Select Database File** under the **Set Up** menu item in the modifications screen to select the REF database. Use the blackboard mode (**<F10>**) to leave the cursor below the Status Symbols line in the form. Use **Load Fields** to add the **MAJ_AUTHOR, SEC_AUTHOR, TITLE, REFERENCE,** and **YEAR** fields to the form. Switch to the blackboard mode to move the fields and add any needed text to complete the form. FORM2 now accepts data from two databases, EXTRACT and REF.

(4) Select **Create a View**, from **Create** in the main menu of the Assistant. The screen should resemble Figure 3.24 after you have selected this option.

(5) In **Create a View,** you select the databases that will be included in the view, and their respective indices. Select the EXTRACT database and the index that you just created. Select the REF database and the index based upon reference numbers.

(6) Having selected the databases and indices, you now must define the linkage between the two. Use the cursor to select **Relate** from the main menu. The first file that you select will start the relation in dBASE and should the many database of the one-to-many relationship. Under the **Relate item**, you will see the names of the databases that you have selected to be in the view. Select the EXTRACT database first. The REF database will be displayed to the right in a separate secondary screen. After selecting the REF database from the secondary screen the Assistant will ask you what field to use for the linkage. Press **<F10>** to see a field list. The field list is from the EXTRACT database. The index field on REF is

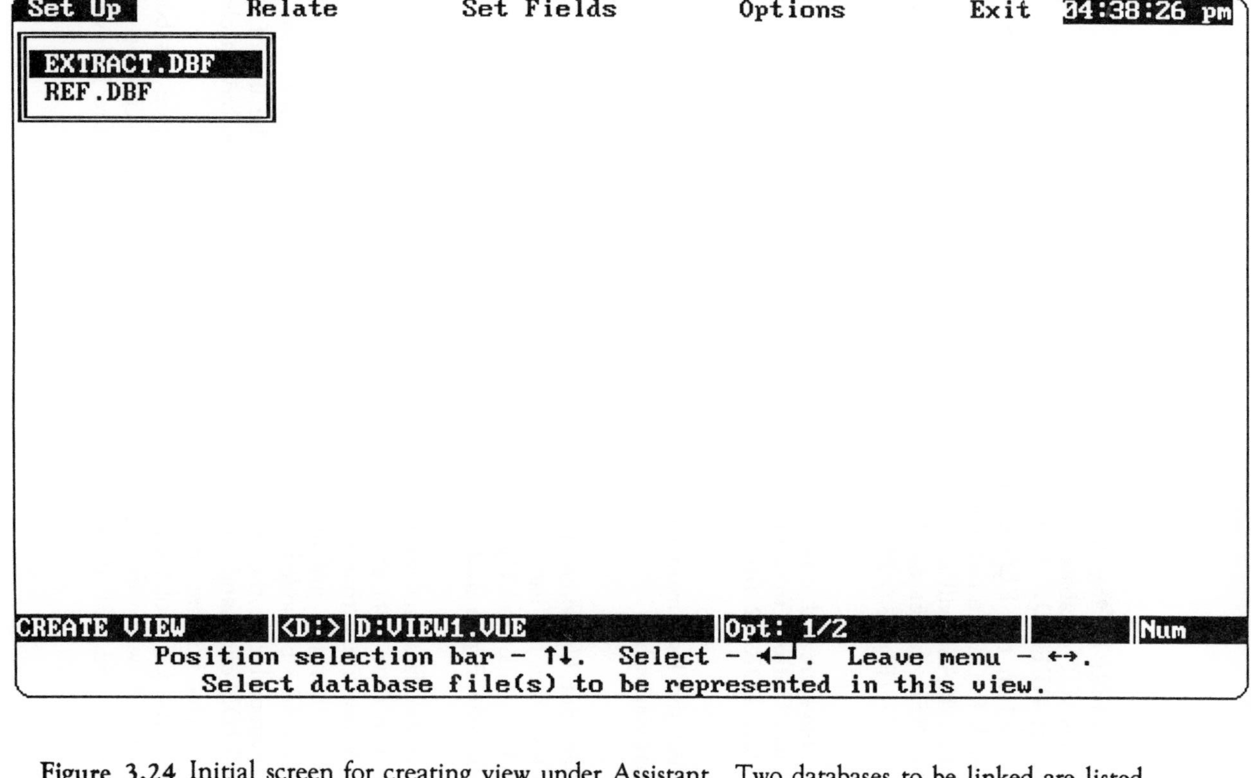

Figure 3.24 Initial screen for creating view under Assistant. Two databases to be linked are listed and can be selected. As each is selected a list of index files is displayed and user selects correct index to match database being used.

NUMBER, so the link field should be the same variable in the EXTRACT database to provide the correct linkage between the two. Select **REFNO.** In this manner, dBASE will know that the value located in the REFNO field in EXTRACT is the value needed to search for in the index file of the REF database. By using the REFNO value, dBASE will use the REF index to locate to correct the record in the REF database. Before pressing **<enter>**, the screen should resemble Figure 3.25.

(7)You have created a view, but now need to tie FORM2 to that structure. Select **Options** from the main menu and select **Format** from the menu. Select **FORM2** from the possible list.

(8)Select the **Exit** main menu item and **Save** your result under any name that you wish.

(9)Select **Set Up** when you return to the Assistant main menu and select **View** from the list. Select the view that you just created and then go to **Edit** under the **Update** main menu item to see the results. It should resemble Figure 3.26.

Once the databases are linked, the linkage is invisible to the user. Each screen can show data that have been retrieved from different sources, but you will not know that necessarily unless you know something about the view file that you are using. Views can be modified and additional databases may be added to the chain. However, spend a great deal of time contemplating the structure that you are building. The use of the ***RELATION*** command quickly can become obtuse unless you have a clear image of what you are linking and how those linkages work.

Using the Control Center of dBASE IV, the view structure has been incorporated with the **Query** file discussed earlier. Linking the EXTRACT and REF databases in this situation involves the following steps:

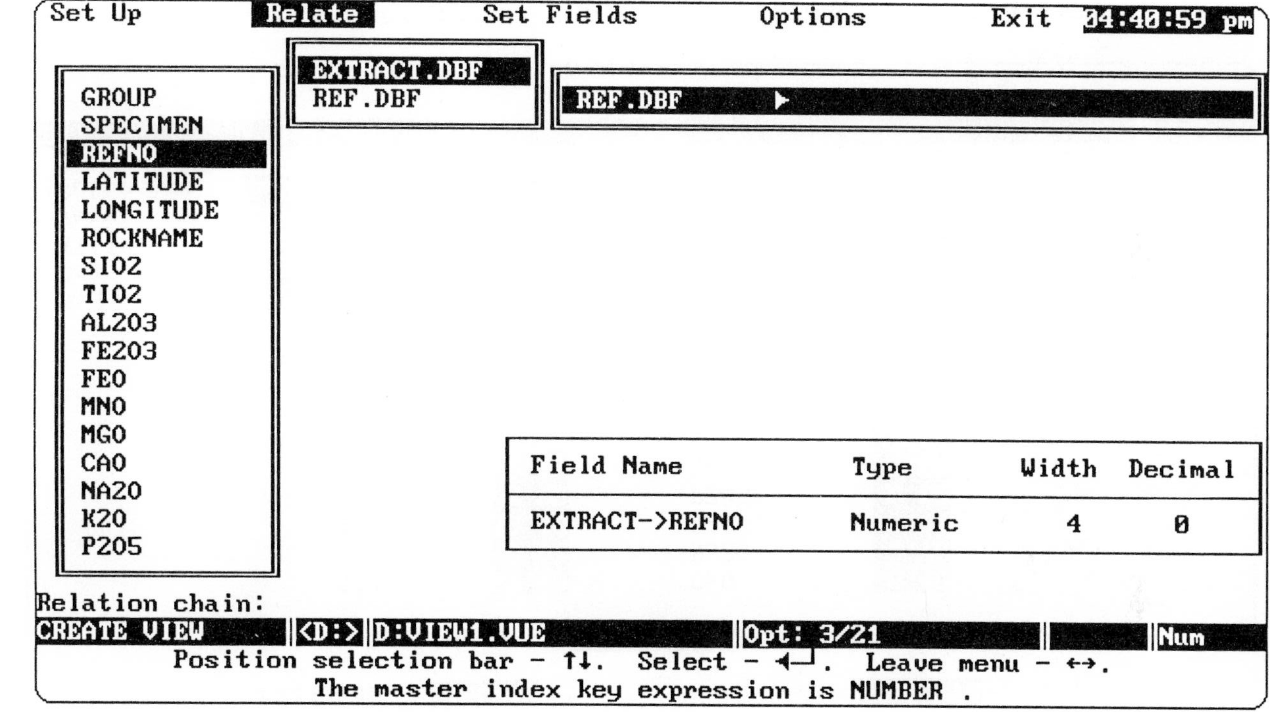

Figure 3.25 Constructing relationship between EXTRACT and REF databases under Assistant.

```
          Group  BTO         Specimen  A        Reference No.   20

               Latitude     15.210          Longitude    145.770

                    Rock Name       ANDESITE

                   SiO2        57.40          CaO           8.48
                   TiO2         0.62          Na2O          2.76
                   Al2O3       18.40          K2O           0.74
                   Fe2O3        4.88          P2O5          0.00
                   FeO          2.26          CO2           0.00
                   MnO          0.00          H20_PLUS      1.30
                   MgO          2.40          H20_MINUS     0.00

         Status_Symbols:  2A,3C,3I

 Maj. Author Taylor, S.R.,
 Sec. Author Capp, A.C.,  Graham, A.L., and Blake, D.H.,
 Title       Trace Element Abundances in Andesites II, Saipan, Bougainville and
 Year        1969
 Reference   Contributions to Mineralogy and Petrology, v.

EDIT         ||<D:>||EXTRACT          ||Rec: 1/408         ||     ||Num
```

Figure 3.26 Combined screen form with data from two sources (EXTRACT and REF) under dBASE III Plus.

(1) Create a form to see the results of the linked files. You may create a new **Form** or **Modify an earlier form** to include all of the fields shown in Figure 3.27. The **Layout** menu item on the screen creation facility has an entry for attaching the screen to an additional database to extract fields from more than one source.

(2) Select create from the **Queries** list in the center of the screen.

(3) Select **Add file to query** from the **Layout** menu item. Select the EXTRACT database and add all of the fields by pressing **<F5>**. Tab to the REFNO field and remove it from the list by pressing **<F5>** while under REFNO.

(4) Select **Layout (<Alt>< L>)** from the main menu and **Add file to Query** to add the REF database to the query file. Select all of the fields in the REF database by pressing **<F5>**.

(5) To create the linkage between the two files move the cursor to the NUMBER field of REF and select **Create link by pointing** from under the **Layout** menu item. LINK1 will be displayed on the REF description. Move to the EXTRACT database by pressing **<F3>**. Move the cursor to the REFNO field and press **<enter>**. The other part of the linkage will be displayed.

(6) If the form and query have been selected, you can look at the data by pressing **<F2>**. The information will be displayed within your screen design and with data from two different databases.

```
Layout  Fields  Words  Go To  Exit                                3:40:22 pm
          Latitude    999.999               Longitude     9999.999

                             Rock Name     XXXXXXXXXXXXXXXXXXXXXXX


                                        SiO2   99.99
                                        TiO2   99.99
                                        Al2O3  99.99
                                        Fe2O3  99.99

                  Status Symbols XXXXXXXXXXXXXXXXXXXXXXXXXXXXXXXXXXXXXXXX

                  Reference XXXXXXXXXXXXXXXXXXXXXXXXXXXXXXXXXXXXXXXXXXXXXXXXXXXXXX
                  Major Author XXXXXXXXXXXXXXXXXXXXXXXXXXXXXXXXXXXXXXXXXXXXXXXXXX
                  Year  9999



Form     ||D:\dbiv\<NEW>           ||Row:14 Col:25   ||File:Ref        ||    Num    Ins
            Add field:F5  Select:F6  Move:F7  Copy:F8  Size:Shift-F7
```

Figure 3.27 Possible form to be used with view under dBASE IV.

3.17 Programming in dBASE

The real power of dBASE is in its programming language. Each operation that the Assistant or Control Center performs can be duplicated from the command line. The ability to write a file of such commands, a program, makes dBASE a malleable system.

The Assistant helps the user to learn the command language of dBASE by building the commands on the line above the status line near the bottom of the screen. As you select items from the menu(s) the different parts of a dBASE command appear on this line. All dBASE commands begin with the dot prompt("."). When in the Assistant try selecting a few items and see what the dot prompt commands for each are. Using the Assistant, select the EXTRACT database using the index created previously. Note the command line before you press **<enter>** after giving the system the name of the index file. It should read "**USE EXTRACT INDEX<index file name>**". Now **Edit** the first record in the database, using the Assistant. The command line should read "**EDIT**" as you strike the **<enter>** key selecting **Edit** from the pull-down menu. Now try the same operations using the command language. Close all of the files from the Assistant. Press **<esc>** from the main menu of the Assistant. You now should have a blank screen with only the status line from the Assistant remaining. dBASE has supplied you already with a dot for a prompt. Type in the first command ("*USE ...*") and press **<enter>**. The program should go to the default disk drive and try to locate the correct file(s). If the command is executed correctly, the status line will show that you are in the database named EXTRACT and positioned at the first record. Then type in the next command line, **EDIT** and you should be at the same location in the database and see the default edit screen as before. By using the Assistant in this manner, you can begin to learn the dBASE command language itself. To reenter the Assistant from the command line, type in the command ***assist***. Notice that nothing has changed in the system, you still are in the same database.

Using the Assistant, recreate each of the searches you have done previously, and record the command line in each case. Using only the command line, perform the searches again. For example, "How many samples have SiO_2>50% and what is their average Al_2O_3 content?" should have created a command line that reads "***AVERAGE FOR SIO2>50 AL2O3*** " in the Assistant just before you pressed **<enter>** for the last time. Compare the results of that operation with being in the command mode and entering that single line. With the persistent use of the Assistant in this manner, creating single line commands for searches becomes easier because you are acquainted already with the necessary phrases and conditions. For each command, many other options that are not included in the Assistant, are described in the voluminous documentation that accompanies dBASE III Plus.

All of the dot prompt commands created by the Assistant work under dBASE IV. In addition, dBASE IV has addition commands and new clauses added to old commands that surpass the abilities of dBASE III Plus.

3.18 INTERPRETING A SIMPLE DBASE PROGRAM

This exercise will acquaint you with the rudiments of dBASE programs that are common to both dBASE III Plus and dBASE IV packages. Rather than create a program, we will use the current example database and interpret a simple program for calculating ratios of major oxide values. The program was written to minimize the types of commands introduced to simplify the structure. Ordinarily, the program would be completely menu-driven and would only ask the user for whatever information is required.

The program that we will be dissecting is listed in Table 3.2. The user selects the appropriate database using dot prompt commands before program execution. The program requires that the fields with the major oxide values be named accordingly, that is SIO2, AL2O3, FEO. EXTRACT has this format, but any other database with these fields could be used. In order to work on a particular database with this program ,we must first select it. This is done using the command line *USE EXTRACT*. The program will add the value of the ratio that we have chosen to the database, so a new field named RATIO must be added to the database itself. This is done via the *MODIFY STRUCTURE* command. The editing menu screen (see Fig. 3.28) then can be adjusted for the new field by adding it at the end of the listing. Modify the structure to add a new field name RATIO as a numeric field. The size of the field should be adjusted for the values that you are creating depending upon the oxides that you will select. When you are finished, leave this screen by pressing <ctrl> < end> to save your work. To run the program, type in "*DO*

Table 3.2 RATIO Program

```
*  Program:     RATIO.PRG
*  Author:      J. FRIZADO
*  Date:        07/21/87
*  Notice:      Copyright (c) 1987, J. FRIZADO, All Rights Reserved
*  Reserved:    den, num, denr, numr, star, m_ratio

*-- Module 1- Set Environment
*  This group of commands sets the programming environment.

SET TALK OFF
SET BELL OFF
SET ESCAPE OFF
SET CONFIRM ON
STORE 0 TO num
STORE 0 TO den
STORE 0 TO numr
STORE 0 TO denr
STORE " "  TO star

* -- Module 2- First Screen
*  This series of commands clears the screen and builds the
*  initial screen.

CLEAR
@2,0 TO 20,79   DOUBLE
@3,26           SAY [R A T I O            P R O G R A M]
@4,1 TO 4,78    DOUBLE
@8,10           SAY [This program uses a previously selected database.]
@9,10           SAY [The database should have fields named for the
major oxides.]
@10,15          SAY [The database should be on the default disk drive.]
@13,10          SAY [The database should be altered to include a blank
"RATIO" field.]
@14,20          SAY [Are you ready to proceed?  (Y/N)]
@14,51          GET star PICTURE "A"
READ
IF STAR<> "Y"
      RETURN
ENDIF
```

Table 3.2 cont'd

```
* -- Module 3 - Choose a numerator
*    This series of commands allows the user to select a numerator
*    from a menu listing.

CLEAR
@2,0 TO 20,79    DOUBLE
@3.26            SAY [R A T I O                    P R O G R A M]
@4,1 TO 4,78     DOUBLE
@6,32            SAY [SELECT A NUMERATOR]
@8,37            SAY [1.          Fe203]
@9,37            SAY [2.          FeO]
@10,37           SAY [3.          MGO]
@11,37           SAY [4.          CAO]
@12,37           SAY [5.          NA2O]
@13,37           SAY [6.          K20]
@14,37           SAY [7.          P205]
@15,37           SAY [8.          Na20+K20]
@16,37           SAY [9.          FeO+Fe203]
@17,37           SAY [0.          exit]
@19,23           SAY " select    "
@19,42           GET num PICTURE "9" RANGE 0,9
READ
If num = 0
        RETURN
ENDIF

*--    Module 4- Choose a denominator
*    This series of commands allows the user to select a denominator
*    from a menu listing.

CLEAR

@2,0 TO 20,79    DOUBLE
@3,26            SAY [R A T I O                    P R O G R A M ]
@4,1 TO 4,78     DOUBLE
@6,32            SAY [SELECT A DENOMINATOR]
@8,37            SAY [1.          SiO2]
@9,37            SAY [2.          A1203]
@10,37           SAY [3.          Ti02]
@11,37           SAY [4.          Fe203]
@17,37           SAY [0.          exit]
```

Table 3.2 cont'd

```
@19,33          SAY  " select   "
@19,42          GET den PICTURE "9" RANGE 0,4
READ
If den = 0
        RETURN
ENDIF

*--   Module 5-Calculations
* This series of commands calculates the ratio and records
* the result in a field named RATIO in the database.

CLEAR
GOTO TOP
DO WHILE .NOT. EOF ( )
   DO CASE
      CASE num = 1
          numr = FE203
      CASE num = 2
          numr = FEO
      CASE num  = 3
          numr = MGO
      CASE num = 4
          numr = CAO
      CASE num = 5
          numr = NA20
      CASE num = 6
          numr =  K20
      CASE num = 7
          numr = P205
      CASE num = 8
          numr = NA20+K20
      CASE num = 9
          numr = FE203+FEO
   ENDCASE
   DO CASE
      CASE den = 1
          denr = SI02
      CASE den = 2
          denr = AL203
      CASE den = 3
          denr = AI02
      CASE den = 4
```

Table 3.2 cont'd

```
        denr = FE203
    ENDCASE
    IF denr <> 0
        m_ratio= numr/denr
        * The following statement will print the result to the
        * screen.
        7 m_ratio
        REPLACE RATIO WITH m-ratio
    ENDIF
    SKIP
ENDDO
RETURN
```

RATIO". The program then will present the user with a few menu screens of selections before altering the entire database by adding the value of the selected ratio to each database record. When the program is finished, it will return to the dot prompt.

The program is divided into several modules. Comments are prefaced by an *. The first module creates the environment for the program to be executed in. The SET commands determine how the dBASE program will respond in certain situations. For example, SET TALK OFF instructs the program not to copy actions to the screen. If talk were set to the on position, then each time the program performed an operation, the screen would display whatever characters were involved. When the program writes 2.35 to the disk, 2.35 would be displayed on the screen. The STORE commands are used to initialize variables. dBASE allows the user to create memory variables. These are values that are stored in memory and not in the database. The STORE command instructs dBASE that the following are memory variables and the initial values to be placed in the reserved memory locations.

The second module creates the first screen that the user sees (see Fig. 3.29). The CLEAR command clears the screen before creating the new display. The SAY command displays a line of text at the screen locations specified by the "@" part of the command. The DOUBLE command draws the double lines that are displayed on the screen. The GET command allows for user input. It causes the program to place the user inputted value into a particular memory variable under a specified format. PICTURE "A" specifies that the input will be a single character. The READ command instructs dBASE to wait on a user inputted value

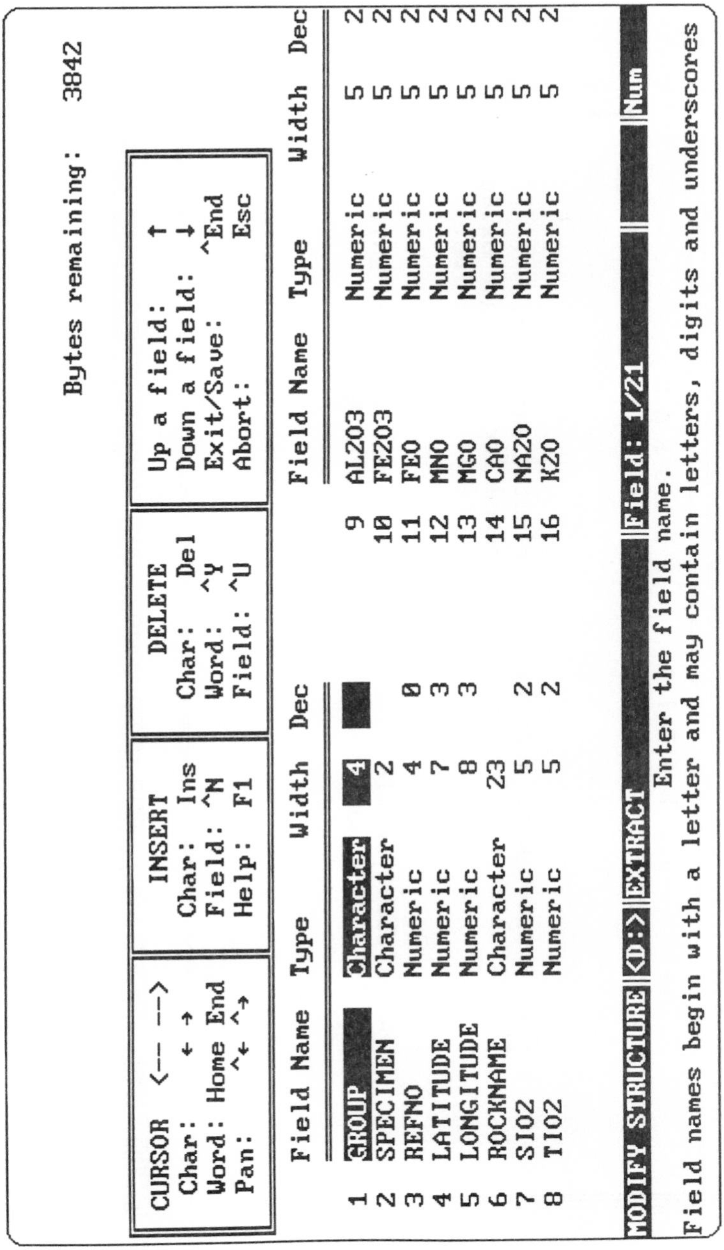

```
                                                  Bytes remaining:   3842

 CURSOR  <-- -->     INSERT         DELETE         Up a field:     ↑
  Char:    ← →      Char:  Ins     Char:   Del     Down a field:  ↓
  Word: Home End    Field: ^N      Word:   ^Y      Exit/Save:     ^End
  Pan:     ^← ^→    Help:  F1      Field:  ^U      Abort:          Esc

     Field Name  Type       Width  Dec           Field Name  Type       Width  Dec

   1 GROUP       Character     4              9  AL2O3       Numeric        5    2
   2 SPECIMEN    Character     2             10  FE2O3       Numeric        5    2
   3 REFNO       Numeric       4    0        11  FEO         Numeric        5    2
   4 LATITUDE    Numeric       7    3        12  MNO         Numeric        5    2
   5 LONGITUDE   Numeric       8    3        13  MGO         Numeric        5    2
   6 ROCKNAME    Character    23             14  CAO         Numeric        5    2
   7 SIO2        Numeric       5    2        15  NA2O        Numeric        5    2
   8 TIO2        Numeric       5    2        16  K2O         Numeric        5    2

MODIFY STRUCTURE |<D:>|EXTRACT          |Field: 1/21  |       |Num
                         Enter the field name.
Field names begin with a letter and may contain letters, digits and underscores
```

Figure 3.28 Modify Structure screen in dBASE III Plus.

```
                    R A T I O     P R O G R A M

        This program uses a previously selected database.
        The database should have fields named for the major oxides.
             The database should be on the default disk drive.

        The database should be altered to include a blank "RATIO" field.
                  Are you ready to proceed?  (Y/N)

Command          <D:> EXTRACT                Rec: 1/408               Num
                      Enter a dBASE III PLUS command.
```

Figure 3.29 First screen of the RATIO program.

and acts in conjunction with the GET statement. The final three lines of code test the user input to determine what to do. The If statement tests the condition that the input is not equivalent to a lowercase "y". If the user inputs a "y", then the condition is false and control is transferred to the line after the ENDIF command which closes this module. If the user presses any other key, then the program terminates by RETURNing to the dot prompt.

The third and fourth modules are similar to the second module. Each one draws a screen format or menu that the user selects a choice from. The differences are only in the GET commands. In these two modules, the PICTURE "9" instructs dBASE that the input will be a single digit and the RANGE part of the command states the acceptable range of values for that input. The test section at the end of each looks at the user input to determine if the user wishes to exit the program.

The fifth module performs all of the operations and the calculation of the ratio itself. GOTO TOP is the same as in the Assistant, it sets the record pointer to the first record in the database. The DO WHILE command at the start of this module is a widely used command in the dBASE language. It is the start of a series of commands that will be performed while the specified condition is met. The first time that the condition is false, the do loop is completed and control reverts to the statement after the ENDDO statement. In this situation, the condition is ".NOT. EOF()". The EOF() function returns a false value until the record pointer tries to read past the end of the database. At that point, the end-of-file indicator changes the value of the function to true. By specifying ".NOT. EOF()", the program will continue until the last record is processed, then control will be transferred to the ENDDO statement. That also is why it was important to take the record pointer to the start of the database before issuing this command. Otherwise, the process would start from wherever the record pointer was positioned and perform the operations until the last record was accessed.

The DO CASE command is a variation of the DO command. In this situation, there is a series of mutually exclusive cases. Each case is a series of commands that will be performed if the conditions are such for that case to be selected. The conditions for each case are specified in the CASE statements. The commands to be performed if the statement is true are listed under that CASE statement and terminated by the next CASE statement or the ENDCASE statement. In this example, the program will be looking at the value of num, which is a memory variable that indicates what the user has selected as the numerator field. If the user has selected

1 as the menu choice from the numerator menu, then num=1 is true and that case will be performed. In this program, the value of the field FE2O3 then will be assigned to the memory variable num. If num =8, then only "CASE num=8" will be operational and the value of the NA2O and K2O fields will be added together and placed into the memory variable num. The second block of Do CASE sets the value of the denominator.

The last section of the fifth module calculates the ratio and inserts it into the database. The first IF statement checks the value of the denominator to determine if a division by zero would occur. If the denominator is zero, the program does not perform any operation but control reverts to a SKIP command. SKIP causes the record pointer in the database to move to the next record. The DO WHILE do loop then is terminated, so control goes back up to the DO WHILE statement which checks to see if the last record has been processed, if not then the process of generating the ratio is performed on the next record. If the denominator is not zero, a new memory variable m_ratio (to differentiate from the field RATIO) is calculated and that value REPLACEs the value in the field named RATIO in the current record. The IF statement block is then ended and the pointer SKIPs to the next record for the process to continue. When the last record has been processed, control transfers to the statement after the ENDDO which returns the user to the dot prompt.

Modifications to the program such as additional ratios and customizing can be performed via ***MODIFY COMMAND.*** You also may use an ASCII-based word processor. If you know that only certain values should be added to your database and you have named them accordingly, then the menu sections of this program can be removed and the calculation section enlarged to accommodate your calculations. For example, Table 3.3 is a short program named AFM that calculates the values for plotting an AFM ternary diagram and inserts those values into fields named A, F, and M in the database.

3.19 USING A MICROCOMPUTER DBMS FOR MANAGING GEOLOGICAL INFORMATION

dBASE is a widely used database management software packages for MS-DOS computers. The number of these computers present in business and business's use of data insure that the microcomputer DBMS market

Table 3.3 AFM program

```
*       Program:    AFM.PRG
*       Author:     J. FRIZADO
*       Date:       07/21/87
*       Notice:     Copyright (c) 1987, J. Frizado, All rights reserved
*       Reserved:   m_alk,m_fe,m_mg,tot

SET TALK OFF
SET BELL OFF
SET ESCAPE OFF
SET CONFIRM ON
STORE 0 TO m_alk
STORE 0 TO m_fe
STORE 0 TO m_mg
STORE 0 TO tot

*   This program requires that a database is selected and the fields
*   for the major oxides are labelled by their names.  It will calculate
*   the co-ordinates for an AFM diagram and place them in the database
*   in fields named A, F, and M.  These fields must be added to the
    database before running this program.

GOTO TOP
DO WHILE .NOT. EOF()
        m_alk=NA2O+K2O
        m_fe=FEO+(0.8999*FE2O3)
        m_mg=MGO
        tot=100/(m_alk+m_fe+m_mg)
        REPLACE A WITH m_alk*tot
        REPLACE F WITH m_fe*tot
        REPLACE M WITH m_mg*tot
        SKIP
ENDDO
RETURN
```

always will be changing. As new versions of software are created and as new versions of dBASE are created, the "tug of war" for market share and dominance continues. For the geologist this means that there are many software products to choose from that will be sufficient for scientific purposes and that many of the "bells and whistles" that sell new database management software to businesses are useless to the geologist. An example is report writing/labeling. The ability to format your output in a variety of ways is important to a scientist, but not the ability to generate various mailing labels.

All of the exercises and structure that have been discussed within this chapter can be mimicked on any number of commercial and shareware products. These principles also apply to DBMS systems on other types of computers. The geologist as consumer needs to make the selection as to what fits within his own personal framework of computer system, finances, and most importantly, what he intends to do with the data.

References

Dinerstein, N., 1987, *dBase III Plus for the programmer:* Scott, Foresman & Co., Glenview, Illinois, 637 p.

Jones, E., 1987, *Using dBase III Plus:* Osborne/McGraw-Hill, New York, 516 p.

Lima, T., 1989, *Inside dBase IV*: Addison-Wesley Publ. Co., New York, 370p.

CHAPTER 4

Use of Spreadsheets for Data Manipulation and Display

Jay Parrish
R.E. Wright Associates, Inc.
Middletown, Pennsylvania USA

4.1 INTRODUCTION TO LOTUS

4.1.1 Features and Requirements

A spreadsheet is similar to a huge piece of gridded paper on which you may make calculations. The advantage of an electronic spreadsheet is that as you enter a value or formula into each gridded block, or cell, the entire spreadsheet may be recalculated. Lotus is a powerful tool for geoscientists. It is possible to purchase templates (prewritten spreadsheet formulae) into which you may import your data for a variety of geological applications. A few of these would include oil well analysis, potential field data reduction, and surveying. It can be useful particularly in constructing mathematical models which include many variables. An advantage of a spreadsheet (over a program written in BASIC or PASCAL) is that a spreadsheet allows you to change values rapidly and view a graph of the result immediately.

The first popular microcomputer-based spreadsheet was Lotus 1-2-3. It always has been expensive (several hundred dollars), and early versions were copy-protected. One of the distinguishing features of Lotus was the use of a bar menu which had many selections available at the touch of a letter while still displaying a screen of information.

Lotus will run on virtually all MS-DOS machines with sufficient memory. Early versions ran on as little as 128K but the most recent release requires 640K for efficient use. This is because the workspace is held in memory and too large a spreadsheet will result in an error message. An 80 column display should be used with 24 or more lines. Lotus is used best with a hard disk but may be used with a two floppy drive system.

Lotus has spawned many imitators. Some, such as Quattro and Excel are comparable in price, but there are clones available for as little as $29. Many of these programs are useful to the casual user. Lotus was written with business applications in mind but has proven to be useful for scientific problems. Lotus 1-2-3 obtained its name because it had three basic features desired by most users: spreadsheet, graphing, and database.

Upon entering Lotus by typing ***Lotus*** you will be presented with a bar menu offering :

1-2-3 PrintGraph Translate Install Exit

by moving the cursor with the arrow keys or typing in the highlighted letter one will invoke that particular program. The primary spreadsheet program is 1-2-3. Once a graphic has been created it may be printed using PrintGraph. Data may converted from and to other spreadsheet formats using Translate. Perhaps one of the most frustrating aspects of using Lotus is installing the program for your system configuration. This can be time consuming as you insert five or more disks and make selections from menus.

Lotus has on-screen help which can be invoked at any time by using the **<F1>** key (see Fig. 4.1). **Help** will be presented for the subject area that you are using currently. You also may go to the **Index** (see Fig. 4.2) and select from a list of topics. Using the **<esc>** key will return you to the spreadsheet. Perhaps one of the most comforting things about Lotus is the

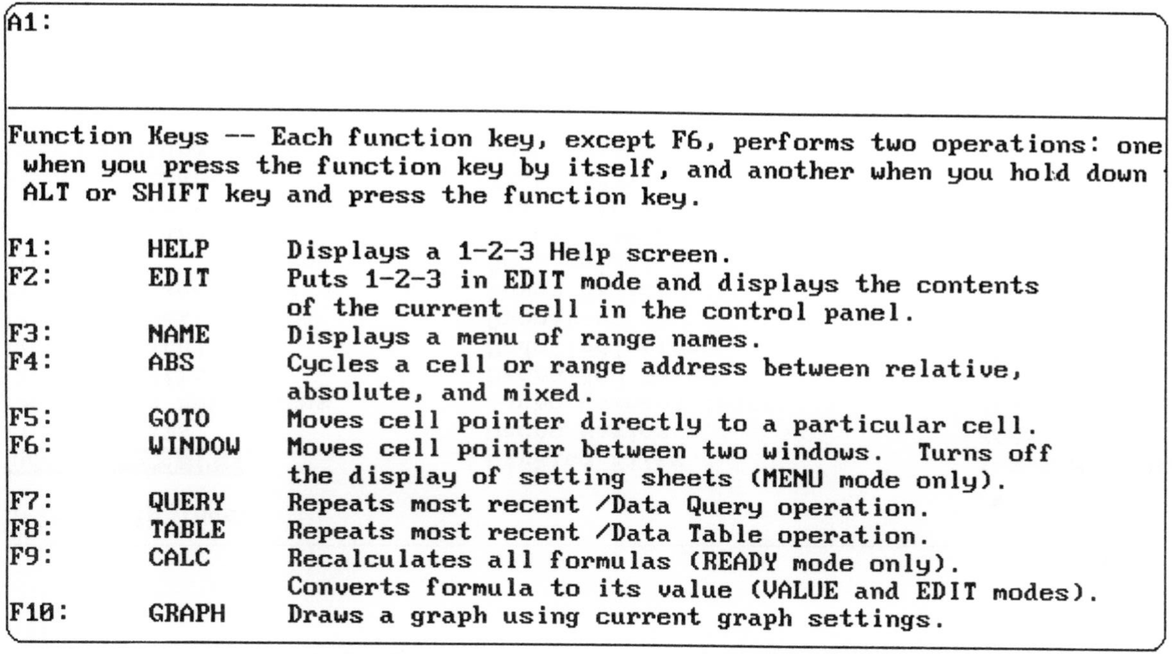

```
A1:

Function Keys -- Each function key, except F6, performs two operations: one
 when you press the function key by itself, and another when you hold down
 ALT or SHIFT key and press the function key.

F1:      HELP     Displays a 1-2-3 Help screen.
F2:      EDIT     Puts 1-2-3 in EDIT mode and displays the contents
                  of the current cell in the control panel.
F3:      NAME     Displays a menu of range names.
F4:      ABS      Cycles a cell or range address between relative,
                  absolute, and mixed.
F5:      GOTO     Moves cell pointer directly to a particular cell.
F6:      WINDOW   Moves cell pointer between two windows.  Turns off
                  the display of setting sheets (MENU mode only).
F7:      QUERY    Repeats most recent /Data Query operation.
F8:      TABLE    Repeats most recent /Data Table operation.
F9:      CALC     Recalculates all formulas (READY mode only).
                  Converts formula to its value (VALUE and EDIT modes).
F10:     GRAPH    Draws a graph using current graph settings.
```

Figure 4.1 Function key screen

```
A1:                                                                  HELP

1-2-3 Help Index

About 1-2-3 Help            Linking Files               1-2-3 Main Menu
Cell Formats                Macro Basics                /Add-In
Cell/Range References       Macro Command Index         /Copy
Column Widths               Macro Key Names             /Data
Control Panel               Mode Indicators             /File
Entering Data               Operators                   /Graph
Error Message Index         Range Basics                /Move
Formulas                    Recalculation               /Print
@Function Index             Specifying Ranges           /Quit
Function Keys               Status Indicators           /Range
Keyboard Index              Task Index                  /System
Learn Feature               Undo Feature                /Worksheet

To select a topic, press a pointer-movement key to highlight the topic and then
press ENTER.  To return to a previous Help screen, press BACKSPACE.  To leave
Help and return to the worksheet, press ESC.
```

Figure 4.2 Help Index screen

esc feature. If you suspect that you are lost in the menus or about to makea mistake, pushing the **<esc>** key will take you back one layer in the menus.

Whatever is currently in the highlighted cell will be displayed in the extreme upper-left corner. For example, upon entering an empty spreadsheet it will display "A1:" with nothing to the right (see Fig. 4.3). After entering a value, for example "13", it will display "A1:13". If you would like to edit a cell, the **<F2>** key puts you in edit mode. If you are entering text, it should be preceded by a "'" for left justified, or a "^" for centered.

Lotus creates ASCII print files and can import print files in ASCII format directly. To import an ASCII file into your spreadsheet, rename the file with the extension .PRN. While in 123 you may import data into an existing database or start a new file. When your cursor is located at the starting point you desire for imported data use **/FI**. This will present you with a directory of possible files to import, which are only those files with the extension .PRN. Note that this means that you may combine a .PRN file with an existing spreadsheet file, being careful not to write over portions of your existing file.

Your next selection is whether it is a Text or Numerical file. If you select text it will import each line as a single cell of text: not numbers. Alternatively, if you import as numbers it will separate your data into lines of cells with each delimited value in a different cell.

If you place the text in quotes it will be entered as a separate cell. However, if you do not have quotes around text you will have to replace text with words in quotes using a locate and replace feature of a word processor.

Rename EXTRACT.CDF as EXTRACT.PRN using the DOS command REN. This replaces the original file with the same file labeled only with a different extension. Alternatively, you may use COPY EXTRACT.CDF EXTRACT.PRN and then have two copies of the file, each with different extensions.

After using **File Import** to bring your ASCII file into the spreadsheet, the data may be viewed either as a database which can be sorted, or numerical data which can be analyzed in a statistical manner, or both. The datafile does not have the names of the fields. Use Table 2.3 to identify each data item. By inserting a row above row 1 and entering the names of the fields, you can mark each column as a particular data entry. When exporting your data file you may make use of the **Translate** portion of Lotus to convert your worksheet format file to a format picked up easily

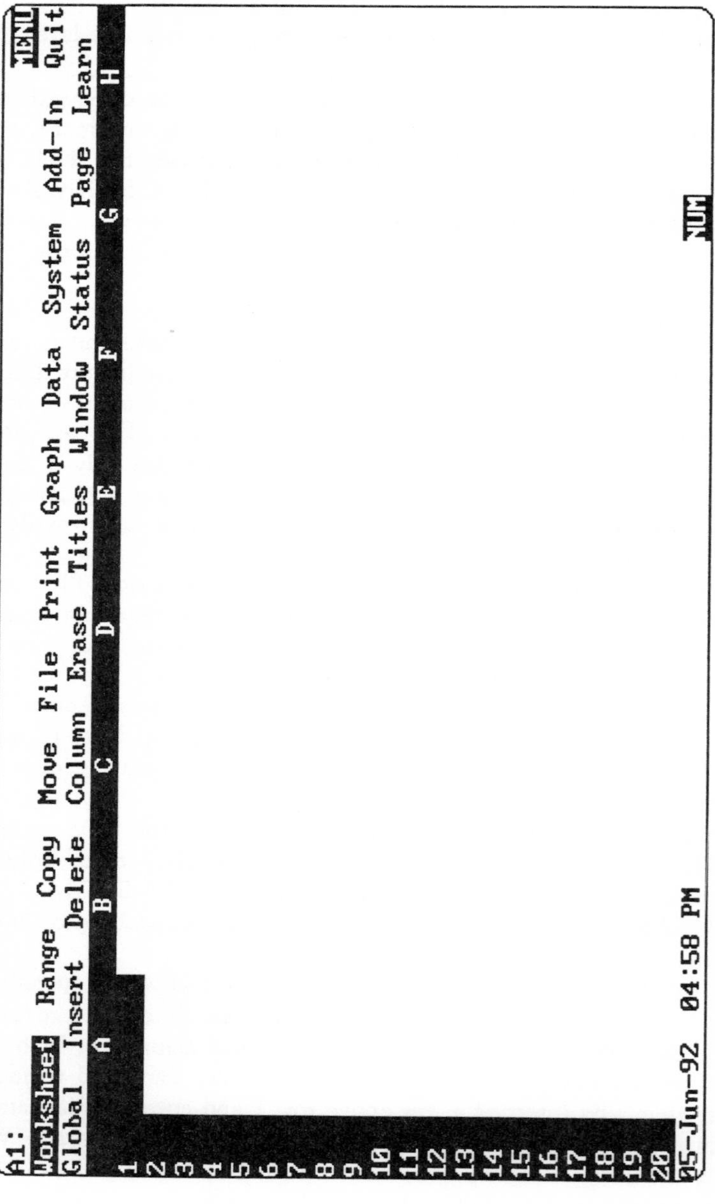

Figure 4.3 Main Menu screen of 123.

by other spreadsheets. Lotus version 1 worksheet files have .WKS extenders while later versions have .WK1 extenders. Any .WKS file will be read by a later version of Lotus and saved as a .WK1 file. If you desire an ASCII file, create a .PRN file (**/Print File**). Under **Options**, note that the number of lines per page may not exceed 100. Therefore your output file for EXTRACT will have four page breaks (a blank line every 100 lines), which can be removed with a word processor.

It is possible to travel quickly from one cell in a spreadsheet to any other point by using the GOTO key (**<F5>**), otherwise you can spend time paging up and down in your database. If your processor is slow, your keystrokes will exceed the screen redisplay and you will end up overshooting your target point. It also is possible to split the screen and keep one portion stationary while allowing roving on the other portion (**/Worksheet Windows** and then select **Horizontal** or **Vertical**).

Each cell has a unique designation ranging from A1 to IV8192. It is possible to write formula in each cell. The end product of the calculation will be placed in the cell containing the formula. Formula should begin with a "+" sign to avoid being mistaken for text. Most usual mathematical and statistical functions are available. They are preceded by an "@" sign. In addition, it is possible to do logical functions. This is useful for marking or sorting out a particular set of values. After entering a value or formula, you may either hit ***<return>***, which will leave you in the cell you were working on, or use a directional arrow to move you to an adjacent cell.

If you desired statistical information on the SiO_2 content of the entire database, you could travel to the last row of data by entering ***<end>*** and the down arrow. The ***<end>*** key in combination with an arrow allows you to travel to the end of any contiguous column or row of data. Below the G column of data type **@AVG(G1..G408)** (see Fig. 4.4). This will take an average of the values in the range from G1 to G408. If you want averages for every column, type **/ Copy** (**/C**), enter return after selecting the cell or cells you want copied, and then mark the cells you would like the formula copied to. The column letter will change automatically as the formula is copied to adjacent columns. If you should ever not want the column or row to change automatically, it is possible to fix a cell value by preceding the column and or the row designation by a "$".

G410: @AVG(G2..G409) READY

	A	B	C	D	E	F	G	H
391	CCW	Y	4369	30	93	BASALT	51.44	0.85
392	CCX	M	4372	25.4	112.5	GRANITE	73.47	0
393	CCX	L	4372	25.4	112.5	GRANITE	71.15	0
394	CCX	K	4372	25.4	112.5	GRANITE	75.89	0
395	CCX	J	4372	26	115	GRANITE	74.88	0.05
396	CCX	I	4372	26	115	GRANITE	74.92	0.05
397	CCX	H	4372	25.5	114.4	GRANITE	73.69	0.04
398	CCX	G	4372	25.5	114.4	GRANITE	73.67	0.04
399	CCX	F	4372	23	110	GRANITE	74.38	0.07
400	CCX	E	4372	26.4	116.3	GRANITE	71.02	0.52
401	CCX	D	4372	25.5	115	GRANITE	72.88	0.17
402	CCX	C	4372	30	118	GRANITE	74.74	0.15
403	CCX	B	4372	30	118	GRANITE	61.56	0.93
404	CCX	A	4372	30	118.4	GRANITE	69.01	0.43
405	CDC	G	4376	44	91	BASALT	50.23	0.47
406	CDC	F	4376	44	91	BASALT	50.86	0.76
407	QX	B	7501	58.25	-155.17	ANDESITE	65.01	0.64
408	QX	A	7501	58.25	-155.17	ANDESITE	65.51	0.56
409	BTO	B	20	15.23	145.75	ANDESITE	57.2	0.54
410							57.57279	

02-Jun-92 09:13 AM UNDO

Figure 4.4 Computing average SiO2 contents of entire EXTRACT database.

4.2 SORTING

Data Sort (/DS) can be used to sort the data according to ascending or descending values. For example, if you wished to view all samples listed from low to high SiO_2 content, you could type **/DS**. Then select Column G as your primary sort field as shown in Figure 4.5. Alternatively, you could sort by rock name (column F) as your primary key and SiO2 as the secondary key (see Fig. 4.6, results as Fig. 4.7). Be sure to sort the entire database, otherwise just the values on which you are sorting will shift and your database will be left a meaningless matrix of numbers with no way to return to the original order. Whenever doing data sorts it always is prudent to save a copy of your spreadsheet in progress. This may be performed by typing **/ File Save (/FS)**. If you have never saved the spreadsheet before you will be given the opportunity to name the file at this pint. If you have previously saved the file most spreadsheets will ask if you want to replace the file or cancel the request.

Data Fill also is a useful function. It may be used to attach a sequential number to a line of data. This will aid you in resorting your database to its original order. Type **/ Data Fill (/DF)** and then highlight the column which will contain the sequential numbers. You also will be asked to provide the range of values and increment. In our situation, we would need to add a column by using **/ Worksheet Insert Column (/WIC)**. This will insert a new (blank) column for your data fill operation.

It is possible to sort the database by rock name, and then perform individualized statistics on each rock type.

/DF allows the user to create a sequence of numbers with some set increment. This is useful for creating a bin range in performing a **Data Distribution (/DD)** or merely numbering your samples.

4.3 RECALCULATION

It is customary to use a spreadsheet in a "what if" mode: what happens to the final result if I change this number? After you have entered a large

```
G410:                                                                   MENU
Data-Range  Primary-Key  Secondary-Key  Reset  Go  Quit
Specify primary order for records
                              Sort Settings
    Data range:          A2..U409

    Primary key:
      Field (column)     G2..G2
      Sort order         Ascending

    Secondary key:
      Field (column)
      Sort order
401 CCX       D          4372     25.5      115 GRANITE     72.88      0.17
402 CCX       C          4372       30      118 GRANITE     74.74      0.15
403 CCX       B          4372       30      118 GRANITE     61.56      0.93
404 CCX       A          4372       30    118.4 GRANITE     69.01      0.43
405 CDC       G          4376       44       91 BASALT      50.23      0.47
406 CDC       F          4376       44       91 BASALT      50.86      0.76
407 QX        B          7501    58.25  -155.17 ANDESITE    65.01      0.64
408 QX        A          7501    58.25  -155.17 ANDESITE    65.51      0.56
409 BTO       B            20    15.23   145.75 ANDESITE     57.2      0.54
410
```

Figure 4.5 EXTRACT database sorted by SiO2 (Column G).

```
A1: 'Group                                                          MENU
Data-Range  Primary-Key  Secondary-Key  Reset  Go  Quit
Specify order for records with same primary key
                            Sort Settings
   Data range:          A2..U409

   Primary key:
     Field (column)     F2..F2
     Sort order         Ascending

   Secondary key:
     Field (column)     G2..G2
     Sort order         Ascending

11 CCM     Q       4358     37      119 BASALT     44.66     2.33
12 DQ      J        206   18.8  -157.05 BASALT     44.77     3.28
13 BXJ     F       3975  12.08  147.062 BASALT      44.8     1.89
14 BWV     A       3961   40.5        9 BASALT     44.93     1.24
15 KD      A        400     43   -114.4 BASALT     45.16     3.88
16 DJ      A        199 -32.99   151.25 BASALT     45.63     2.04
17 FM      A        260   41.8   -121.5 BASALT        46     0.74
18 CAG     AJ      3853  53.77  -168.04 BASALT        46      1.4
19 NN      G        504   19.2     72.8 BASALT     46.05     2.01
20 BTJ     J       1889   64.1    -19.7 BASALT     46.21     3.54
```

Figure 4.6 Sort screen to sort EXTRACT database on Rock Name and secondarily on SiO2.

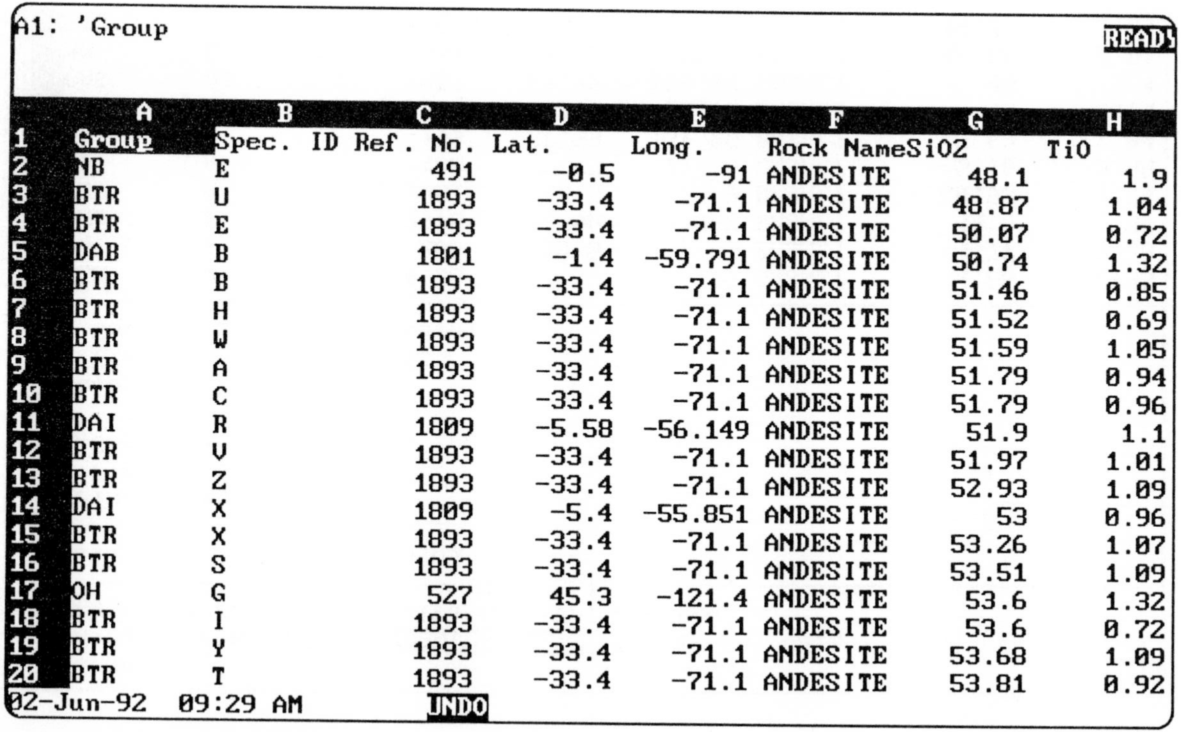

A1: 'Group READY

	A	B	C	D	E	F	G	H
1	Group	Spec. ID	Ref. No.	Lat.	Long.	Rock Name	SiO2	TiO
2	NB	E	491	-0.5	-91	ANDESITE	48.1	1.9
3	BTR	U	1893	-33.4	-71.1	ANDESITE	48.87	1.04
4	BTR	E	1893	-33.4	-71.1	ANDESITE	50.07	0.72
5	DAB	B	1801	-1.4	-59.791	ANDESITE	50.74	1.32
6	BTR	B	1893	-33.4	-71.1	ANDESITE	51.46	0.85
7	BTR	H	1893	-33.4	-71.1	ANDESITE	51.52	0.69
8	BTR	W	1893	-33.4	-71.1	ANDESITE	51.59	1.05
9	BTR	A	1893	-33.4	-71.1	ANDESITE	51.79	0.94
10	BTR	C	1893	-33.4	-71.1	ANDESITE	51.79	0.96
11	DAI	R	1809	-5.58	-56.149	ANDESITE	51.9	1.1
12	BTR	V	1893	-33.4	-71.1	ANDESITE	51.97	1.01
13	BTR	Z	1893	-33.4	-71.1	ANDESITE	52.93	1.09
14	DAI	X	1809	-5.4	-55.851	ANDESITE	53	0.96
15	BTR	X	1893	-33.4	-71.1	ANDESITE	53.26	1.07
16	BTR	S	1893	-33.4	-71.1	ANDESITE	53.51	1.09
17	OH	G	527	45.3	-121.4	ANDESITE	53.6	1.32
18	BTR	I	1893	-33.4	-71.1	ANDESITE	53.6	0.72
19	BTR	Y	1893	-33.4	-71.1	ANDESITE	53.68	1.09
20	BTR	T	1893	-33.4	-71.1	ANDESITE	53.81	0.92

02-Jun-92 09:29 AM UNDO

Figure 4.7 Results of the sort from Figure 4.6.

spreadsheet you may find that every modification you make requires longer and longer to accomplish such that the simple addition of a cell will take several seconds to be displayed on the screen. This is because the entire spreadsheet is being recalculated each time you make a modification. If you select **Global Recalculation Manual (/GRM)** no recalculations will be made until you press **<F9>**. After a modification has been made, a **CALC** message will be displayed in the lower portion of the screen, reminding you to recalculate if you wish to see an updated database.

4.4 GRAPH

Displaying your data in graphical form is accomplished using **/Graph**. Once in Graph there are several options for how you would like to set up your graph: bar graph, stacked bar graph, pie, line (see Fig. 4.8). Frequency distributions are seen best using a bar graph. Trends are best revealed as xy plots or line graphs.

After typing **X** you will be able to define the range of values which will be used as the x axis. Type a period (.) at the starting point to anchor your range. As you extend the range of interest the cells will be highlighted. Upon reaching the last point, press ***<return>*** and you will be back at the **Graph** menu. If, after doing the same for your variables (using A B C D E F) you decide that you would like to remove one range, select **Reset**. You can reset either one or all of the data ranges defined already. Figure 4.9 depicts the graph screen after selections have been made to create a line graph of CaO, TiO_2, and Al_2O_3 as functions of SiO_2 for 50 andesites from the extract file (see Fig. 4.10 for the result). If you have exited from **Graph** and would like to see a graph you made recently, the **<F10>** key brings up the last graph viewed. If you have made a change to a data value, this will allow you to see quickly the effect that the change has had.

Comparisons of two variables are seen best using a line graph. If you would like a scatter plot it is possible to remove the connecting lines by using **Options Format (OF)**.

You may view up to six variables at one time. It is easier to see if you select **Options Color (OC)**. **Choosing View (V)** displays the graph. If the x axis has too many numbers on it, they display on top of each other.

```
A1: 'Group                                                           MENU
Type  X  A  B  C  D  E  F  Reset  View  Save  Options  Name  Group  Quit
Line  Bar  XY  Stack-Bar  Pie
                         Graph Settings
 Type: Line                Titles: First
                                   Second
 X:                                X axis
 A:                                Y axis
 B:
 C:                                            Y scale:      X scale:
 D:                                Scaling     Automatic     Automatic
 E:                                Lower
 F:                                Upper
                                   Format      (G)           (G)
 Grid: None         Color: No      Indicator   Yes           Yes

    Legend:              Format:   Data labels:              Skip: 1
 A                       Both
 B                       Both
 C                       Both
 D                       Both
 E                       Both
 F                       Both
02-Jun-92  09:30 AM
```

Figure 4.8 Graph menu.

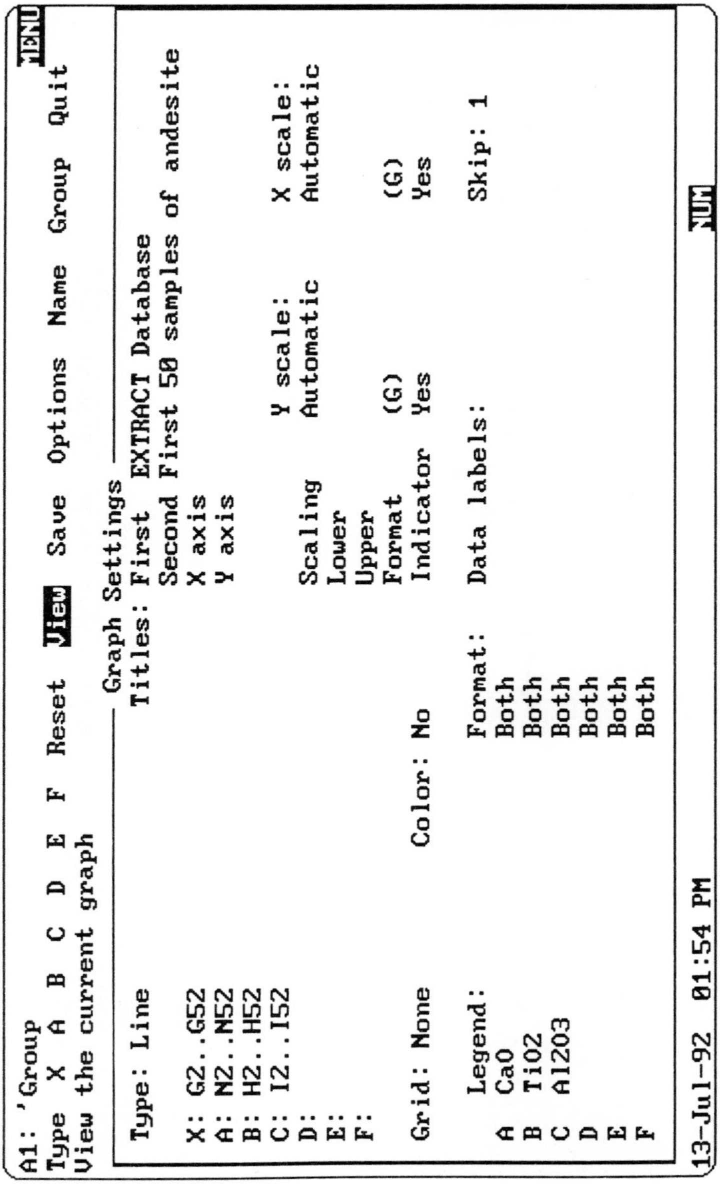

```
A1: 'Group                                                              MENU
Type  X  A  B  C  D  E  F  Reset  View  Save  Options  Name  Group  Quit
View the current graph
                          Graph Settings
 Type: Line                        Titles: First  EXTRACT Database
                                           Second First 50 samples of andesite
 X: G2..G52                                X axis
 A: N2..N52                                Y axis
 B: H2..H52
 C: I2..I52                                          Y scale:     X scale:
 D:                                        Scaling   Automatic    Automatic
 E:                                        Lower
 F:                                        Upper
                                           Format    (G)          (G)
 Grid: None          Color: No             Indicator Yes          Yes

    Legend:                     Format:    Data labels:           Skip: 1
 A  CaO                         Both
 B  TiO2                        Both
 C  Al2O3                       Both
 D                              Both
 E                              Both
 F                              Both

13-Jul-92  01:54 PM                                          NUM
```

Figure 4.9 Graph menu. with entries for CaO, TiO2, and Al2O3 vs SiO2 graph.

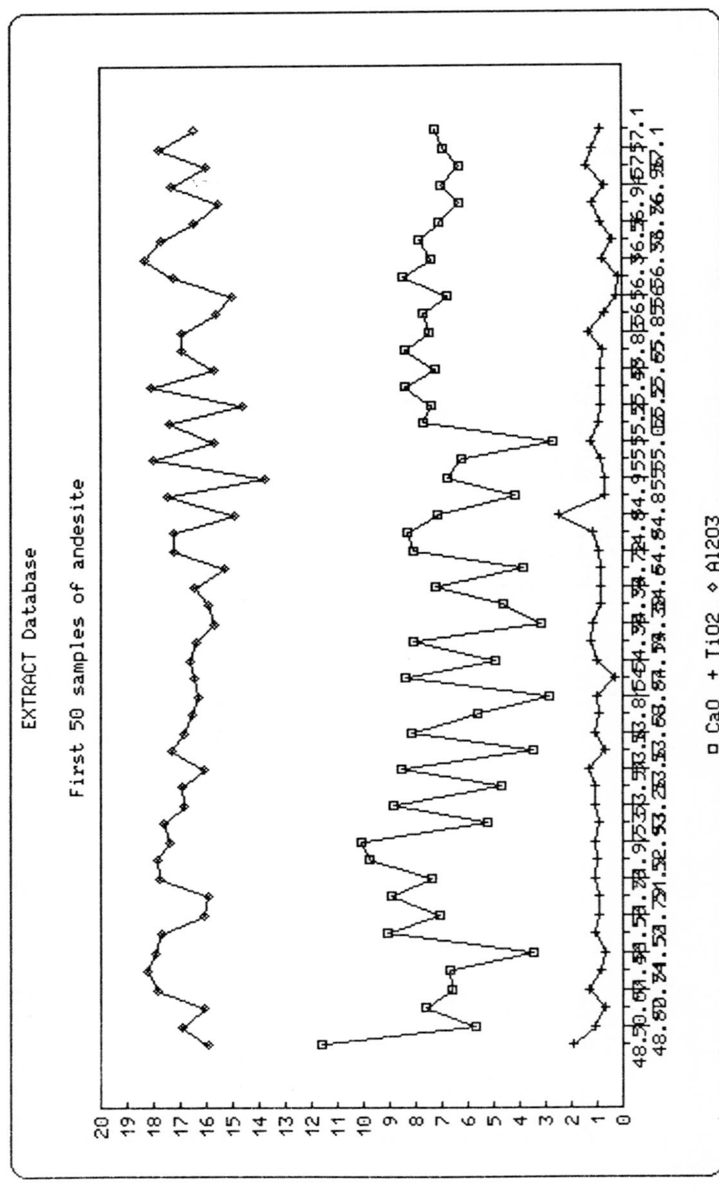

Figure 4.10 Graph of CaO, TiO_2, and Al_2O_3 vs SiO_2 for 50 samples from EXTRACT database.

To avoid this, use **Options Scale Skip X-axis (OSSX)** and then enter a skip factor. This will display every nth value. Choosing **Options Title (OT)** brings you to labeling options. You may label the x and y axes and create two lines of titles running across the top of your graph. It also is possible to create a legend to make it easier to discriminate between several data ranges. Always use **Name Create (NC)** to save a graph within the spreadsheet, otherwise,upon leaving the spreadsheet the graph is lost.

Using the **Save** command you will create a print file which may be sent to a printer after you have exited the spreadsheet program and entered the **PrintGraph** program. An alternative is to invoke GRAPHICS.COM in your system set-up and then use the **<Printscreen>** to make a copy while still in the spreadsheet. This has the disadvantage of a long wait while the screen is dumped to the printer.

4.5 PRINTGRAPH

Upon leaving the spreadsheet by answering Yes after **Exit**, you may select **PrintGraph** from the main menu and will see the screen as shown in Figure 4.10. In **PrintGraph** you define the print parameters for your hardcopy. On the right-hand side of the screen is a listing of hardware and software settings. You may alter the default directory to your data disk or change the output printer. You may also select one half-page or full-page output, and fonts for titles. By selecting **Select** you will be able to mark all the files in the default directory which you would like to print. This is done with the space bar. A second press deselects the file. After a batch of files have been selected, advance one page (to make sure you are aligned) and then type **Go (G)**. **PrintGraph** is slow and you may determine that this is something you do before stepping out for lunch.

This brief discussion of how to use a spreadsheet cannot begin to cover the many applications and tricks of the trade that you will discover with use. Be creative and make plenty of back-up copies!

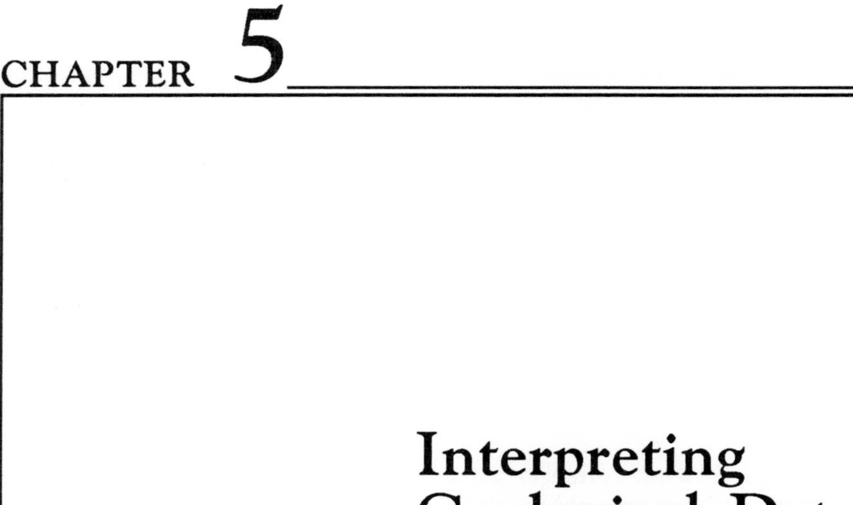

CHAPTER 5

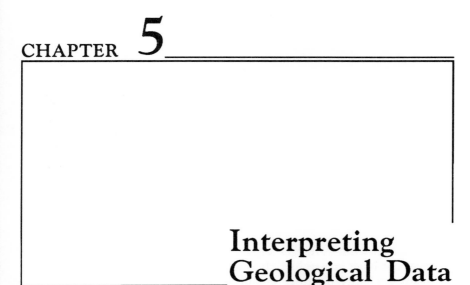

Interpreting Geological Data

R.W. Le Maitre
Dept. of Geology
University of Melbourne
Melbourne, Victoria Australia

5.1 INTRODUCTION

So far this book has dealt mainly with how to create and manipulate databases. However, in most research situations the extraction of relevant information from a database is only the start – you still have to interpret the data. This chapter, therefore, will explain a few of the methods that may be used to examine these data sets.

One of the characteristics of geological data is that it is *multivariate*, in other words each object is characterized by many variables for example a chemical analysis, a set of trace element data, or measurements on fossils etc. The techniques discussed in this chapter, therefore, will be confined to some of the more generally used multivariate statistical methods. As each method has its strengths and weaknesses it is important to understand what they are designed to do before using them. Briefly the methods discussed are:-

(1) *Principal Components Analysis and Factor Analysis* – both are related closely and designed to examine the relationships between objects in a set of data and to explain some of the causes of the variations among the data. Principal components analysis probably is the best method to use to "view" your data in multidimensional space.

(2) *Multiple Discriminant Analysis* – designed to test if predefined groups of data can be distinguished from each other on a basis of their variables.

(3) *Cluster Analysis* – arranges any set of objects into groups, which may or may not have any reality. It is up to the user to justify which of the many similarity measures should be used.

In a nutshell if one is interested in comparing the relationship between *individual* objects then principal components analysis or factor analysis should be used. However, if one is interested in comparing *groups of data* for similarities and differences then multiple discriminant analysis or cluster analysis should be used.

The amount of mathematics in this chapter has been purposely kept to a minimum and wherever possible explanations are given in terms of geometry. The most important mathematic concept to master at this stage is that just as the composition of a feldspar can be represented by a single point in the 3-dimensional space defined by the axes Or, Ab, and An, so a chemical analysis of n oxides can be represented by a single point in the n-dimensional space defined by the n oxide axes. Fortunately, if the concept of n-dimensional space seems difficult to grasp, n can be thought of as 3 in most of the techniques used in this chapter.

5.2 PRINCIPAL COMPONENTS ANALYSIS AND FACTOR ANALYSIS

These two methods are related closely and it is worth spending time to understand the differences. Factor analysis is the oldest method (Spearman,1904) and was proposed to explain the relationships between certain postulated factors involved in educational psychology. However, there is still disagreement as to the meaning of "factor analysis" as it has been widened by many researchers to include other techniques, such as principal components analysis which, unlike factor analysis, assumes no well-defined mathematical model.

Basically one type of factor analysis, termed *R-mode factor analysis* and frequent in the geological literature, sets up a distinct mathematical model which tries to explain the correlations between the *observed* variables in terms of a linear combination of a *specified* number of *unobservable* variables, named *factors*. Mathematically this problem cannot be solved without making further assumptions, which leads to a multiplicity of possible answers. The other type of factor analysis used in the geological literature is *extended Q-mode factor analysis*, popularized by Miesch (1976), which can be thought of as a type of multispecimen mixing model (Le Maitre, 1982) and will not be discussed further.

Principal components analysis (PCA), on the other hand, transforms the original data into a new set, termed principal component coordinates (or scores, to use factor analysis jargon), by means of a rigid rotation of the original axes onto a new set of directions termed *eigenvectors*. Each successive eigenvector defines a direction of decreasing amount of "spread" in the data. Associated with each eigenvector is an *eigenvalue* which is a sum of squares (or variance) and a measure of the amount of "spread" of data along that direction.

As most of the examples of "factor analysis" in the geological literature are of R-mode type which, as usually used in its simplest form, is no more than principal components analysis, only the latter method will be discussed further. However, if any readers still wish to perform factor analysis it is recommended that they should read some of the comments on the two methods found in Blackith and Reyment (1971), Cattell (1965),

Francis (1978), Jöreskog, Klovan, and Reyment (1976), and Le Maitre (1982).

5.3 DETAILS OF PRINCIPAL COMPONENTS ANALYSIS

There are two basic methods of performing PCA; one involves extracting the eigenvectors and eigenvalues from a variance-covariance matrix, the other from a correlation matrix.

In the *variance-covariance matrix* each diagonal term is the variance of the variable corresponding to the row (or column) of the matrix. The off-diagonal terms are covariances between the two variables corresponding to the row and column of the matrix. For example, in Table 5.2, 257.088 is the variance of the variable HVV for the *Fusispirifer* group and 100.795 is the covariance between the variables T and HDV. Note that the matrix is square and symmetrical.

In the *correlation matrix*, each diagonal term is 1 and the off-diagonal terms are correlation coefficients between the two variables corresponding to the row and column of the matrix. For example, in Table 5.4, the correlation coefficient between the variables T and HDV is 0.905. Note that the correlation matrix can be derived easily from the variance-covariance matrix as the correlation coefficient between variable i and j is the covariance between i and j divided by the square root of the product of the variance of i and the variance of j, that is:

$$0.905 = \frac{152.853}{\sqrt{162.999 \times 175.006}}$$

using the variances and covariance from Table 5.3. This relationship is, of course, why the diagonal terms of the correlation matrix are 1.

As the eigenvectors and eigenvalues obtained from the two matrices are different the problem arises as to which one to use. The answer is subjective but, as a general rule, if all the data are measured on the same scale, and are similar in magnitude, then use the variance-covariance matrix. However, if the variables are measured on different scales, for example weights, lengths, ages, concentrations, etc., then the correlation matrix must be used as this gives equal importance to all variables. Under other circumstances it is up to the user to justify the method chosen. Fortunately, it is always possible to decide which method was used if it has not been mentioned, as the sum of the eigenvalues is equal to the sum of the diagonal terms of the matrix from which they were extracted. For example, the sum of the eigenvalues will be equal to the number of variables used if extracted from a correlation matrix. Unfortunately some software packages, particularly those doing "factor analysis", only give the user the choice of using the correlation matrix.

Note that if the raw data is scaled initially so that the value of the ith variable, x_i, is replaced by the standardized variable z_i, where:-

$$z_i = \frac{(x_i - \bar{x})}{s_x} \tag{1}$$

where $\bar{x}$ and s_x are the average and standard deviation of all the xs, then it is immaterial which method is used, as both the variance-covariance and correlation matrices will be identical – because the transformation makes each variable have zero mean and unit standard deviation. This indicates that a program which performs only PCA with the variance-covariance matrix always can be made to use the correlation matrix, by transforming the data using equation #1. Unfortunately the reverse in not true.

By definition the first eigenvector is the direction in which there is maximum "spread" of the data in n-dimensional space; the second eigenvector is the direction in which there is maximum "spread" of the data at right angles to the first eigenvector; the third eigenvector is the direction in which there is maximum "spread" of the data at right angles to the first and second eigenvectors, etc. This is illustrated in Figure 5.1 with a simple 3-dimensional example. Inspection of the eigenvalues, therefore, gives a good indication of how many "factors" are causing the variation in the data. For example, if the first two eigenvalues are large and all the rest are small, then two main causes of variation would be suspected, and their eigenvectors would be used to interpret them.

If the data have a constant sum, such as chemical analyses, the matrices will be singular and the last eigenvalue will be zero. and all the terms of the last eigenvector will be $1/\sqrt{n}$, where n is the number of variables. This simply is a reflection of the fact that all the data lie in a reduced (n-1) space. It also is why you can plot triangular diagrams of what appear to be three variables in 2-dimensional space - given two of the variables the third is known.

In PCA the eigenvectors always are given in normalized form that is the sum of squares of the individual terms is 1. This indicates that each term

Figure 5.1. Diagram to illustrate geometric interpretation of principal components analysis. In 3-dimensional space defined by original axes (variables) X_1, X_2 and X_3, cluster of data points will provide three eigenvectors MV_1, MV_2 and MV_3, where M is mean of all the points. First eigenvector, MV_1, defines direction of maximum "spread" of data; second eigenvector, MV_2, defines direction of maximum "spread" of data at right angles to first eigenvector; and third eigenvector, MV_3, defines direction of maximum "spread" of data at right angles to first and second eigenvectors. Note that eigenvectors always are at right angles to each other.

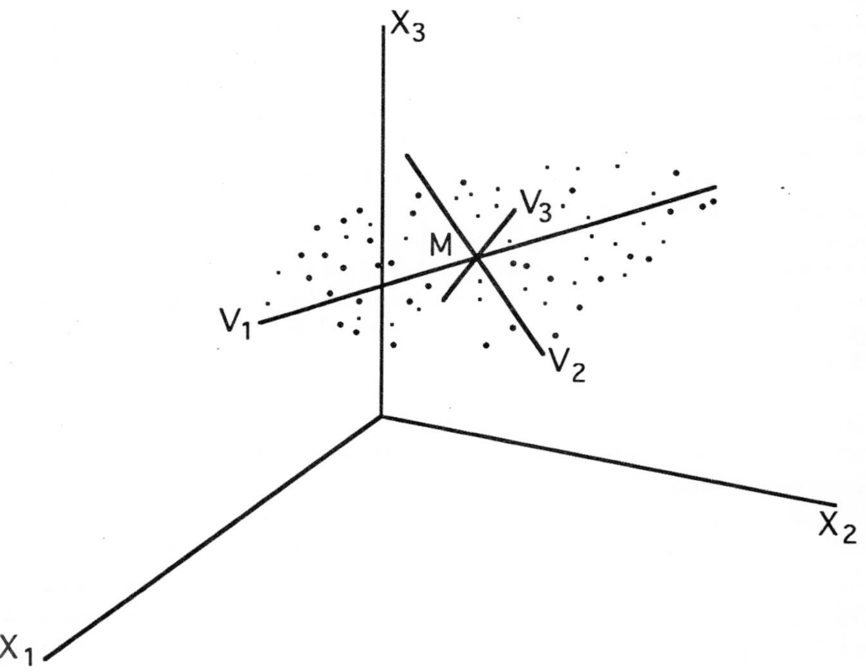

is the cosine of the angle between the direction of the eigenvector and the corresponding original axis (remember the cosine of 90° is 0 and the cosine of 0° is 1). This means that the magnitudes of the individual terms give an indication of the direction (or cause) of the variation. For example, a large value of a particular term indicates that the corresponding variable is important in causing the variation. Similarly, two large terms of equal magnitude, but opposite sign, indicate that they are both important, but that as one increases the other decreases. Note, however, that in R-mode factor analysis the eigenvectors usually are weighted by multiplying each term by the square-root of the corresponding eigenvalue, which makes their interpretation more difficult.

Finally, it is strongly recommended that the data is projected onto the first and second eigenvectors (e.g. Figs. 5.3 and 5.4) in order to see what the data look like when viewed from the two directions which, by definition, display maximum amount of variation in the data for any 2-dimensional projection that can be produced. This is done by calculating the *principal component coordinates* along the required eigenvectors as follows. If the eigenvectors have been extracted from a variance-covariance matrix then c_i, the ith principal component coordinate, is given by the cross product:

$$c_i = \sum_{j=1}^{n} x_{ij} v_j \tag{2}$$

where x_{ij} is the jth variable of the ith set of data and v_j is the jth term of the eigenvector onto which the projection is to be made. However, if the eigenvectors have been extracted from a correlation matrix then it is important to use the following expression:

$$c_i = \sum_{j=1}^{n} z_{ij} v_j \tag{3}$$

where z_{ij} is the jth standardized variable of the ith set of data, as defined in Equation (1). Failure to use the correct equation can lead to misleading results (see Le Maitre, p.114, 1982).

In summary PCA using the variance-covariance matrix produces a simple rotation of axes in the original n-dimensional data space; using the correlation matrix produces a rescaling along each axis, followed by a rotation in the new rescaled n-dimensional data space.

For further details of the mathematics see Le Maitre (1982) who also gives other examples and many other references.

5.3.1 Example 1

This example of principal components analysis uses some measurements taken from the Permian brachiopod *Fusispirifer* (see Fig. 5.2), kindly supplied by Dr. N.W. Archbold of the Department of Geology, University of Melbourne and shown in Table 5.1. As all the measurements are on the same scale (millimeters) and are of the same order of magnitude it was decided to extract the eigenvectors and eigenvalues from the variance-covariance matrix. The results of this calculation together with the principal component coordinates are given in Table 5.2.

The first thing to notice is that the first eigenvector accounts for the vast majority of the variability within the data - nearly 97% of the variance, while the second accounts for nearly all the rest - 2.94%. This means that the original 4-dimensional data can be projected into 2-dimensions with little loss of information. Such a projection is shown in Figure 5.3, using the first and second principal component coordinates given in Table 5.2.

The first eigenvector can be interpreted as indicating physical size, as all its terms are of the same sign, which indicates that they tend to increase (or decrease) together – the larger fossils plotting to the right (e.g. 1 and 2) and the smaller to the left (e.g. 19, 20, 21, and 22).

Figure 5.2. Diagram to illustrate the locations of the four measurements taken on three sets of Permian brachiopods and given in Table 5.1. MW = maximum width; HVV = height of ventral valve; HDV = height of dorsal valve; T = thickness.

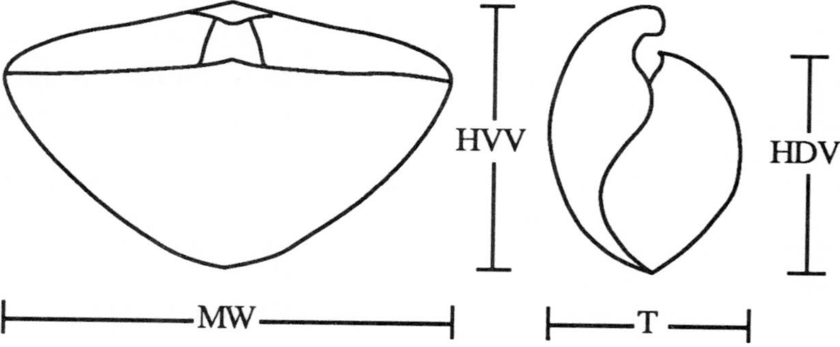

Table 5.1. Raw data used to illustrate methods of principal components analysis, multiple discriminant analysis, and cluster analysis. Data are of three genera of Permian brachiopods (*Fusispirifer*, *Crassispirifer* and *Neospirifer*), kindly supplied by Dr. N.W. Archbold of the Department of Geology, University of Melbourne. Abbreviations: MW = maximum width; HVV = height of ventral valve; HDV = height of dorsal valve; T = thickness; locations of these measurements, which are all in millimeters, are shown in Figure 5.2

		Fusispirifer		
	M W	HVV	HDV	T
1	162.0	50.0	38.5	27.0
2	166.0	64.0	48.5	36.0
3	132.0	49.0	39.5	27.0
4	140.0	56.0	43.0	27.0
5	120.0	49.0	37.0	27.0
6	96.0	35.0	27.0	20.0
7	139.0	43.0	35.0	29.0
8	121.0	40.0	31.0	27.0
9	82.0	24.0	20.0	13.0
10	96.0	35.0	27.0	20.0
11	58.0	21.0	18.0	13.0
12	91.0	43.0	32.0	24.0
13	96.0	48.0	37.0	25.0
14	120.0	47.0	37.0	27.0
15	86.0	49.0	38.0	26.0
16	80.0	44.0	34.0	25.0
17	68.0	20.0	18.0	11.0
18	94.0	42.0	31.0	20.0
19	33.0	16.0	13.0	10.0
20	21.0	9.3	8.4	5.0
21	30.0	10.0	8.5	5.5
22	30.0	11.0	10.0	7.0
		Crassispirifer		
	M W	HVV	HDV	T
1	65.0	34.0	28.0	22.0

Table 5.1 cont'd

2	64.0	21.0	18.0	12.0
3	75.0	46.0	41.0	35.0
4	92.0	57.0	50.0	37.0
5	93.0	51.0	45.0	31.0
6	82.0	33.0	27.0	23.0
7	60.0	21.0	19.0	15.0
8	82.0	32.0	26.0	23.0
9	82.0	53.0	43.0	36.0
10	76.0	36.0	31.0	25.0
11	72.0	52.0	44.0	33.0
12	94.0	49.0	42.0	37.0
13	88.0	57.0	49.0	38.0
14	78.0	45.0	37.0	30.0
15	133.0	71.0	59.0	53.0
16	114.0	70.0	59.0	51.0
17	107.0	73.0	60.0	53.0
18	105.0	47.0	34.0	39.0
19	111.0	55.0	43.0	37.0
20	105.0	44.0	37.0	32.0
21	98.0	45.0	36.0	35.0
22	110.0	52.0	41.0	37.0
23	111.0	67.0	55.0	44.0

Neospirifer

	M W	HVV	HDV	T
1	67.5	54.0	44.0	41.0
2	22.0	16.3	14.5	7.0
3	71.0	49.0	44.0	43.0
4	79.0	54.0	48.0	53.0
5	89.0	60.0	55.0	43.0
6	73.0	59.0	50.0	50.0
7	61.0	51.0	43.0	36.0
8	69.0	56.0	47.0	32.0
9	22.0	17.5	15.0	11.0
10	59.0	41.0	39.0	45.0
11	66.0	53.0	45.0	50.0
12	36.0	21.0	20.0	16.0
13	58.0	39.0	36.0	36.0
14	81.0	46.0	42.0	45.0

Table 5.1 cont'd

15	73.0	49.0	43.0	46.0
16	64.0	57.0	46.0	49.0
17	64.0	35.0	35.0	48.0
18	68.0	41.0	38.0	38.0

Table 5.2. Results of principal components analysis of *Fusispirifer* data from Table 5.1, using variance-covariance matrix. Abbreviations: Raw = actual value of eigenvalue; Sq.Rt. = square root of eigenvalue; As % = eigenvalue as a percentage; Acc. % = accumulative percentage of eigenvalues. For abbreviations of the variable names see Table 5.1 and Fig. 5.2

Variance-Covariance matrix

	M W	HVV	HDV	T
M W	1767.180	607.111	451.700	337.746
HVV	607.111	257.088	188.500	136.693
HDV	451.700	188.500	138.970	100.795
T	337.746	136.693	100.795	76.678

Eigenvectors

	1st	2nd	3rd	4th
M W	0.899	-0.436	0.033	0.013
HVV	0.321	0.702	0.309	0.556
HDV	0.239	0.479	0.157	-0.830
T	0.177	0.296	-0.937	0.044

Eigenvalues

	1st	2nd	3rd	4th
Raw	2170.71	65.91	2.88	0.42
Sq.Rt.	46.59	8.12	1.70	0.65
As %	96.91	2.94	0.13	0.02
Acc. %	96.91	99.85	99.98	100.00

Table 5.2 cont'd

	Principal Component coordinates			
	1st	2nd	3rd	4th
1	175.67	-9.10	1.54	-0.86
2	187.74	6.44	-0.86	-0.93
3	148.62	3.76	0.40	-2.64
4	158.89	6.86	3.38	-1.55
5	137.23	7.79	-0.39	-0.72
6	107.53	1.57	-0.52	-0.82
7	152.26	-5.07	-3.80	-2.06
8	133.81	-1.84	-4.08	-0.73
9	88.50	-5.48	1.08	-1.62
10	107.53	1.57	-0.52	-0.82
11	65.49	1.92	-0.95	-1.94
12	107.51	12.94	-1.17	-0.41
13	114.98	16.96	0.38	-1.67
14	136.59	6.39	-1.01	-1.83
15	106.73	22.80	-0.42	-2.03
16	98.60	19.69	-1.85	-1.62
17	73.80	-3.73	0.94	-2.45
18	108.94	9.27	2.21	-0.28
19	39.68	6.03	-1.30	-1.03
20	24.76	2.88	0.20	-1.31
21	33.19	-0.36	0.26	-0.86
22	34.13	1.50	-0.60	-1.49

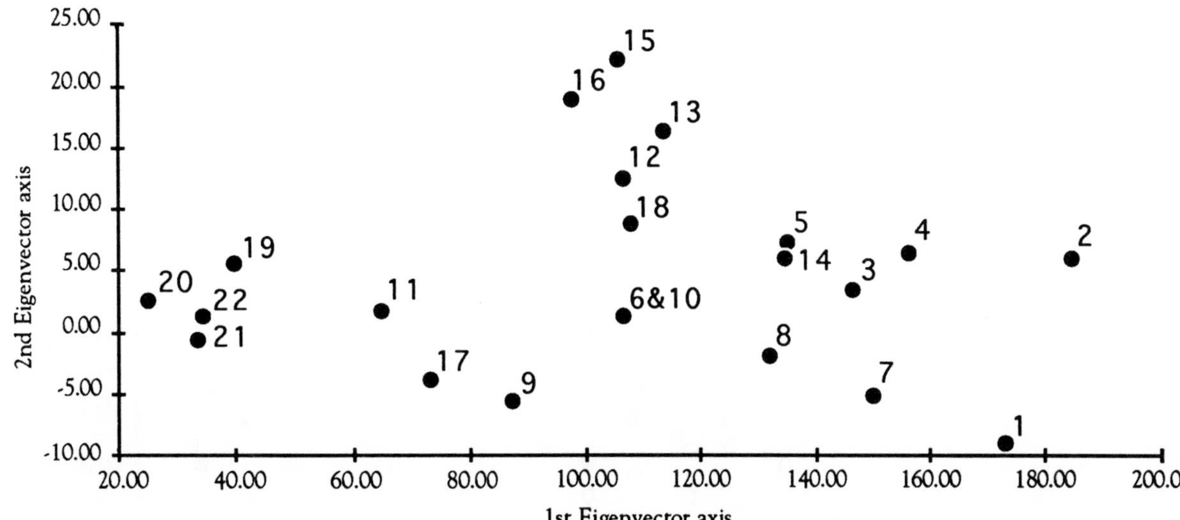

Figure 5.3 Projection of the Fusispirifer data (Table 5.1) onto first two eigenvectors derived from a variance-covariance matrix (Table 5.2) obtained by plotting first vs. second principal component coordinates. Major variation along direction of first eigenvector indicates growth, smallest fossils being to left and largest to right. Minor variation along direction of second eigenvector basically indicates change in shape; more elongate fossils being at top and wider at bottom.

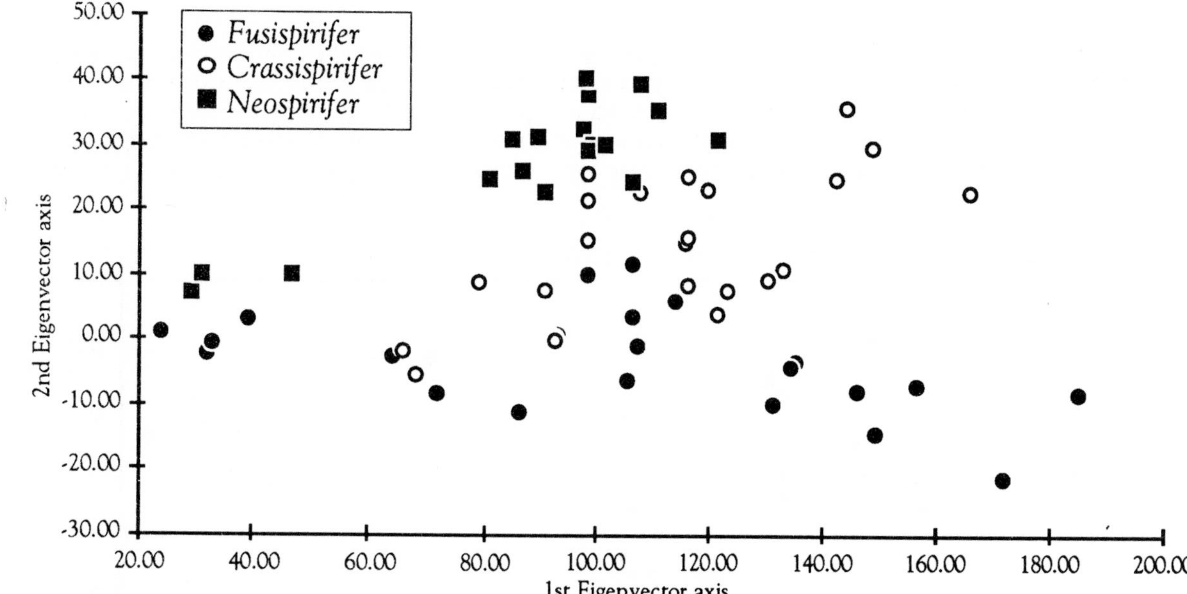

Figure 5.4 Projection of all data from Table 5.1 onto first two eigenvectors derived from variance-covariance matrix (Table 5.3) obtained by plotting the first vs. second principal component coordinates. As with Figure 5.3 major variation along direction of first eigenvector indicates growth, smallest fossils being to left and largest to right. Minor variation along direction of second eigenvector basically indicates change in shape; more elongate fossils being at top and wider at bottom. As can be seen Neospirifer and Crassispirifer tend to be more elongated than Fusispirifer, whereas Fusispirifer shows largest range of sizes, this latter reflecting range of ontogenetic stages of specimens available for measurement (i.e. juvenile, submature and mature).

Table 5.3. Results of principal components analysis of all three genera of Permian brachiopods from Table 5.1, using variance-covariance matrix. For abbreviations see Tables 5.1 and 5.2

Variance-Covariance matrix

	M W	HVV	HDV	T
M W	1005.115	327.228	232.731	160.269
HVV	327.228	239.843	194.112	174.868
HDV	232.731	194.112	162.999	152.853
T	160.269	174.868	152.853	175.006

Eigenvectors

	1st	2nd	3rd	4th
M W	0.859	-0.496	-0.124	0.030
HVV	0.366	0.462	0.547	-0.595
HDV	0.278	0.458	0.290	0.793
T	0.225	0.575	-0.776	-0.128

Eigenvalues

	1st	2nd	3rd	4th
Raw	1261.85	299.00	20.34	1.77
Sq.Rt.	35.52	17.29	4.51	1.33
As %	79.71	18.89	1.29	0.11
Acc. %	79.71	98.60	99.89	100.00

Principal Component coordinates for each group

	Fusispirifer		*Crassispirifer*		*Neospirifer*	
1	174.24	-24.09	81.01	8.94	99.20	35.20
2	187.60	-9.86	70.37	-6.90	30.47	7.28
3	148.38	-9.22	100.53	22.96	100.83	32.30
4	158.79	-8.35	122.12	24.88	112.89	38.22
5	137.38	-4.41	118.04	15.87	123.38	33.49
6	107.28	-7.58	95.20	0.17	109.45	42.70
7	151.39	-16.37	67.88	-2.73	91.12	33.70
8	133.27	-11.81	94.55	-0.76	100.03	31.57
9	87.71	-12.95	109.89	24.21	31.95	10.37
10	107.28	-7.58	92.70	7.51	86.65	33.42
11	65.44	-3.35	100.54	27.44	99.85	41.11
12	108.20	3.19	118.68	16.53	47.77	10.21

Table 5.3 cont'd

13	115.94	5.88	118.63	26.98	82.20	26.44
14	136.64	-5.34	100.51	16.30	108.22	26.19
15	108.22	12.34	168.56	24.33	102.95	32.57
16	99.90	10.60	151.42	32.14	99.65	43.83
17	73.21	-9.92	147.24	38.61	88.32	28.06
18	109.24	-1.52	125.62	7.63	92.53	24.47
19	40.07	2.73	135.76	11.32		
20	24.90	0.60	123.79	3.59		
21	33.03	-3.20	118.54	8.80		
22	34.15	-1.19	133.25	9.52		
23			145.06	26.39		

Table 5.4. Results of principal components analysis of all three genera of Permian brachiopods from Table 5.1, using correlation matrix. For abbreviations see Tables 5.1 and 5.2

	Correlation matrix			
	M W	HVV	HDV	T
M W	1.000	0.666	0.575	0.382
HVV	0.666	1.000	0.982	0.854
HDV	0.575	0.982	1.000	0.905
T	0.382	0.854	0.905	1.000
	Eigenvectors			
	1st	2nd	3rd	4th
M W	0.391	-0.866	0.302	0.075
HVV	0.549	0.017	-0.495	-0.674
HDV	0.545	0.177	-0.378	0.727
T	0.499	0.467	0.722	-0.112
	Eigenvalues			
	1st	2nd	3rd	4th
Raw	3.23	0.66	0.10	0.01
Sq.Rt.	1.80	0.82	0.32	0.10
As %	80.68	16.60	2.49	0.23
Acc. %	80.68	97.28	99.77	100.00

In the second eigenvector, the sign of the term corresponding to MW is negative while all the others are positive. This variation can be interpreted as indicating a change of basic shape, the more elongate (smaller width to height ratio) at the top and the wider (larger width to height ratio) at the bottom.

5.3.2 Example 2

In this example all the data from the three genera of Permian brachiopods in Table 5.1 were used. The interpretation of the eigenvalues and eigenvectors is similar to Example 1, but in this case some generalizations can be made about the three genera. Firstly, both *Neospirifer* and *Fusispirifer*, being nearer the top of the plot, tend to be more elongated than *Crassispirifer*, which tends to occur along the bottom of the plot. In terms of size, *Neospirifer* tends to be smaller than *Fusispirifer*, being more to the left of the plot, while *Crassispirifer* shows the greatest range of sizes.

Note that the three groups of data overlap considerably and that one might, therefore, infer that the groups are not distinct numerically. This would be an incorrect deduction as PCA is not designed to separate groups. The correct method to use to see if the three groups can be separated numerically is multiple discriminant analysis (see Example 3). This is the only reason for including this example here.

5.4 MULTIPLE DISCRIMINANT ANALYSIS

Multiple Discriminant Analysis (MDA) is extremely useful for comparing several groups of data for similarities and differences. For example, are rocks from area A similar to those from area B; can I distinguish three groups of fossils from one another?

With this method one has to be able to assign each set of observations to one of several groups by using criteria that are not involved in the calculation. For example, it is no good trying to see if a group of basalts can be distinguished from a group of andesites, on the basis of their major element chemistry, if they have already been classified using the criteria: basalt SiO_2 < 52%, andesite SiO_2 > 52%!

Basically MDA maximizes the differences between the groups while minimizing the variation within the groups. The results are given in terms of a number of discriminant functions (or eigenvectors) which separate successively the groups from each other by decreasing amounts. Similar to PCA, each eigenvector is associated with an eigenvalue which, if certain assumptions are made about the raw data, can be used for testing the statistical significance of the separations – a feature used rarely in geological applications. Unlike PCA, however, the eigenvectors produced by MDA generally are not at right angles to each other. As the mathematics are more complex than for PCA the reader is referred to Le Maitre (1982) for further details.

The number of discriminant functions always is the smaller of the number of groups minus 1 and the number of variables. So that if, for example, one is trying to separate three groups of fossils on four measurements there would be two discriminant functions, but with six groups there would only be four discriminant functions.

Geometrically each eigenvector defines the normal to a hyperplane (an n-dimensional plane) that "best" separates the groups. The effectiveness of the separation can be judged by noting how many of each group lie on each side of the hyperplane, once the location of the hyperplane is fixed – remember that all the eigenvector indicates to us is the orientation of the plane, not the location. The location can be determined exactly if one makes certain assumptions about the data, or can be determined empirically by looking at the data and selecting a value which minimizes the amount of misclassification. In most geological applications the latter is often the safest method to use particularly if no significance tests are to be applied.

Once it has been decided that the groups can be separated in space, then the discriminant functions can be used to classify other data, assuming that the data do belong to one of the groups. If this assumption is not valid the method, if used simplistically, always will classify any set of data into one of the groups. For example, a sandstone would classify as a basalt or an andesite, if the discriminant function that separates basalts from andesites was applied stupidly to a sandstone.

The method of classifying objects by seeing where an object is positioned along the discriminant functions has a drawback. With two groups and one discriminant function the are two possible outcomes. With three groups and two discriminant functions there are four possible outcomes and with p discriminant functions there are $2^{(p-1)}$, which is more than the number of groups, so that one must look at the data to see where the groups fall in order to come up with a consistent classification strategy. If one has more than one discriminant function, therefore, it is best to project the original data onto the first two discriminant functions to produce a plot, as in Figure 5.5. The discriminant coordinates (or scores) are calculated by using Equation 2. Note, however, that unlike PCA, there is no equivalent to using the correlation matrix, unless the data are standardized before use by using Equation 1 in which situation Equation 2 still is valid. Using such a plot one then is no longer constrained simply to consider the location of the hyperplanes, which would give a vertical and horizontal line on the plot and divide it into four regions, but can define irregular areas around the groups. Theoretically this could also be done in 3-dimensional space with three discriminant functions.

Finally, a word of warning. If the data are singular (e.g. constant sum chemical analyses) the ordinary algorithm for solving MDA will not work as the matrix it has to invert will be singular. For two ways of overcoming this problem see Le Maitre (p.149, 1982).

5.4.1 Example 3

As an example of multiple discriminant analysis, the three genera of Permian fossils from Table 5.1 are used.

The first discriminant function separates cleanly the fusispirifers, which all have negative discriminant function scores, from the neospirifers, which are all positive. However, the second discriminant function does not, by itself, make much further distinction between the groups.

However, if one looks at a plot of the data projected onto the first two discriminant functions (Fig. 5.5) one can delimit areas which give a reasonable separation of the three genera. If required, this procedure could then be refined by noting that the four *Fusispirifer* and three *Neospirifer* at the top of the diagram are all juveniles and could be omitted from the calculation.

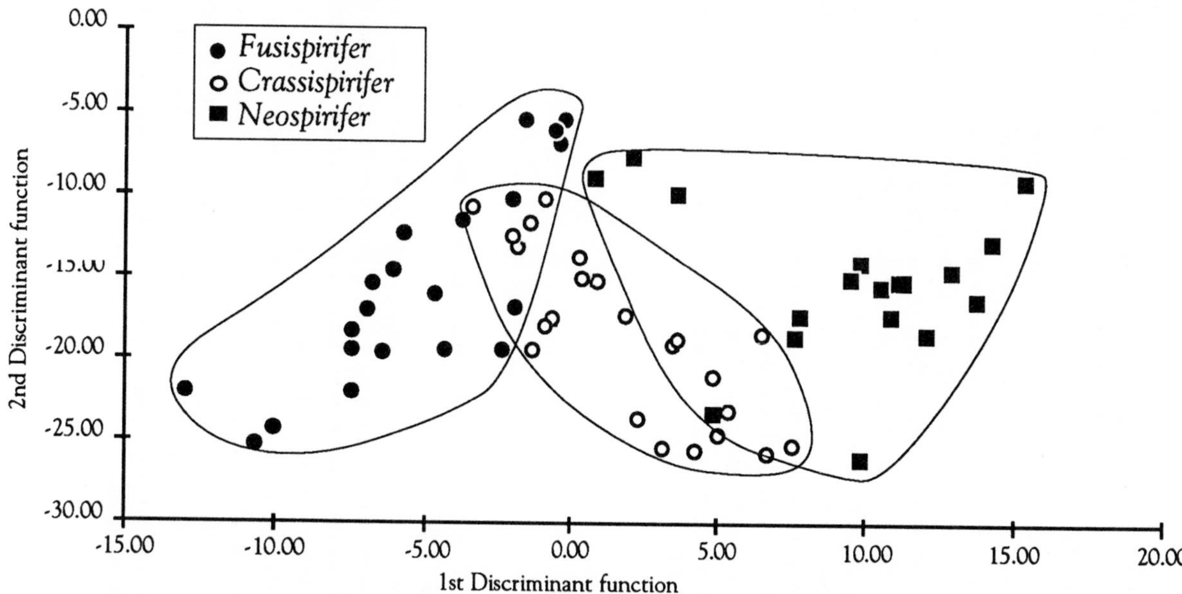

Figure 5.5 Projection of three genera of Permian brachiopods of Table 5.1 into 2-dimensional space of first two discriminant functions given in Table 5.5. Note that unlike Figure 5.4, where there was considerable overlap of three genera, there now is a much clearer separation between genera - in particular Fusisprifer and Neospirifer are completely separate.

Table 5.5. Results of multiple discriminant analysis of three genera of Permian brachiopods from Table 5.1. For abbreviations see Tables 5.1 and 5.2.

Mean group values

	M W	HVV	HDV	T
Fusispirifer	93.7	36.6	28.7	20.5
Crassispirifer	91.2	48.3	40.2	33.8
Neospirifer	62.4	44.4	39.1	38.3

Discriminant Functions

	1st	2nd
M W	-0.145	-0.035
HVV	-0.560	0.145
HDV	0.699	-0.898
T	0.419	0.414

Discriminant Scores for each group

	Fusispirifer		*Crassispirifer*		*Neospirifer*	
1	-13.27	-21.82	0.33	-13.38	7.91	-17.07
2	-10.92	-25.18	-3.43	-10.39	0.75	-8.53
3	-7.66	-21.81	6.69	-18.28	11.04	-17.09
4	-10.29	-24.22	5.19	-24.54	14.06	-16.10
5	-7.66	-19.14	2.40	-23.44	9.96	-26.00
6	-6.27	-14.25	-1.86	-12.81	12.28	-18.20
7	-7.62	-18.05	-0.89	-9.91	7.74	-18.45
8	-6.96	-15.10	-2.00	-12.06	4.90	-23.25
9	-5.90	-11.97	3.57	-18.90	2.10	-7.15
10	-6.27	-14.25	0.96	-14.93	14.60	-12.51
11	-2.14	-9.77	5.02	-20.83	13.16	-14.34
12	-4.85	-15.75	3.79	-18.58	3.70	-9.55
13	-4.46	-19.28	5.49	-23.09	10.00	-13.80
14	-6.54	-19.43	1.92	-17.01	10.71	-15.25
15	-2.45	-19.27	4.40	-25.40	11.31	-15.02
16	-2.00	-16.60	6.88	-25.71	11.49	-15.00
17	-3.87	-11.09	7.75	-25.10	15.70	-8.72
18	-7.10	-16.76	-1.44	-11.25	9.66	-14.83
19	-0.47	-6.37	-1.34	-19.21		
20	-0.29	-4.86	-0.59	-17.27		

Table 5.5 cont'd

21	-1.70	-4.96	0.42	-14.74
22	-0.59	-5.54	-0.91	-17.81
23			3.27	-25.34

5.5 CLUSTER ANALYSIS

This method is designed to investigate groupings within a set of data and, unlike MDA, assumes no prior knowledge of the way in which the objects are related to each other. It also makes no assumptions about the data.

Basically it links successive objects into groups according to the values of certain similarity measures between them and usually presents the results in the form of a *dendrogram*, as shown in Figures 5.6 and 5.7. In both of these diagrams the objects are linked at a particular similarity level by a horizontal line and the nearer the line is to the bottom of the dendrogram, the more similar are the objects. Of course, one disadvantage of such a diagram is that the groupings which occur in multi-dimensional space are represented finally in a one-dimensional diagram.

Before implementing the method the user has to select and justify which of the several *linkage methods* and many *similarity measures* are to be used. Firstly there are three general linkage methods in use:

1. *Single-linkage* – in which an object is linked to a group if it has the closest similarity with *any individual* in the group.

2. *Unweighted average* - in which an object is linked to a group if it has the closest similarity with the *average* similarity measure of the group. Once placed in a group a new average is formed from all the members of the group. This indicates that a new object placed into a large group has little effect on the average.

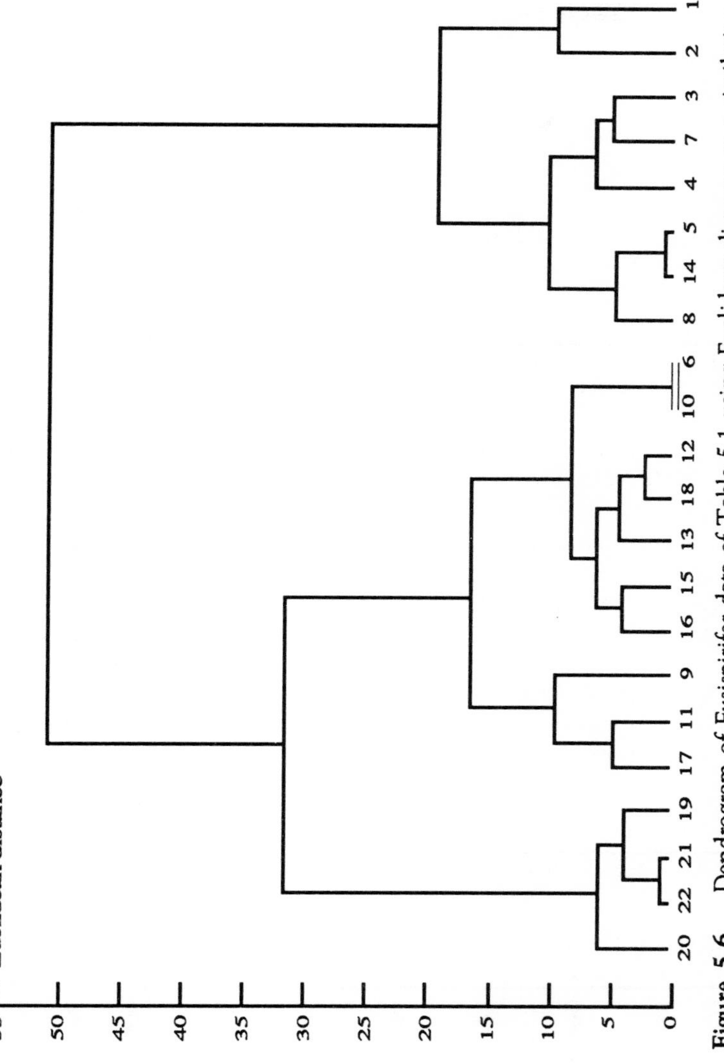

Figure 5.6 Dendrogram of Fusispirifer data of Table 5.1 using Euclidean distance as a similarity measure. Numbers along bottom of dendrogram are same as those used in Table 5.1 to identify each individual specimen. Note the close similarity between these groupings and those of Figure 5.3.

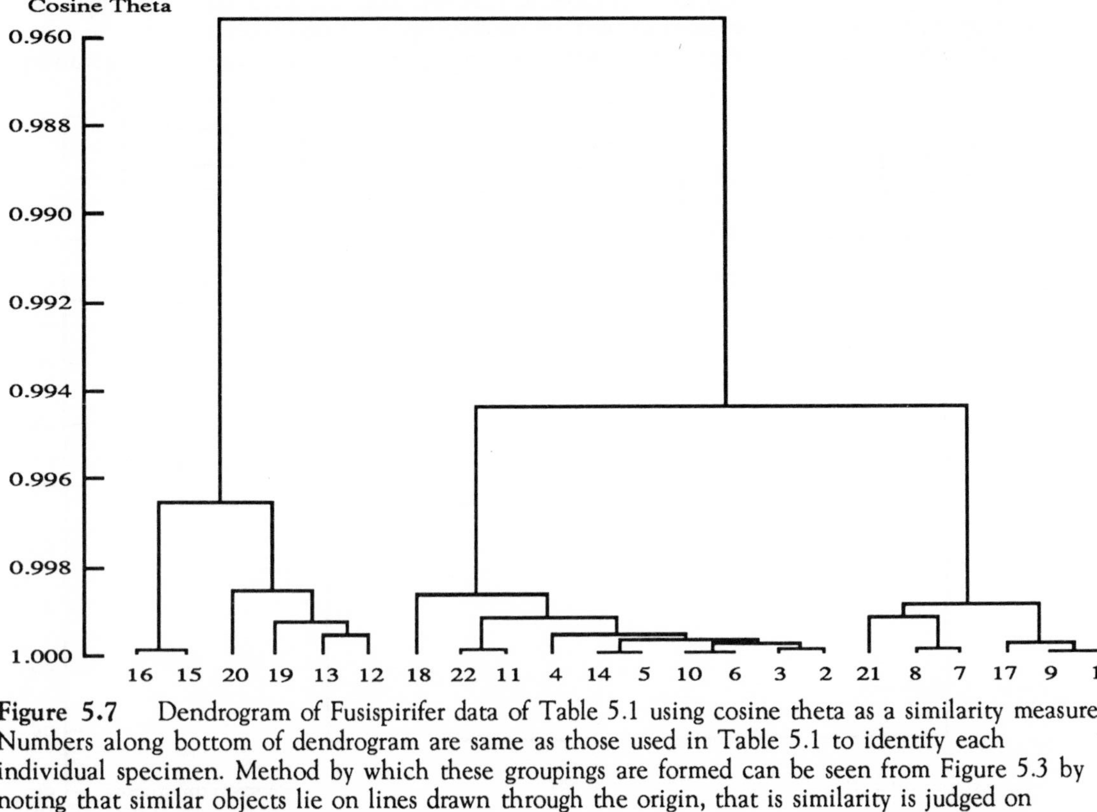

Figure 5.7 Dendrogram of Fusispirifer data of Table 5.1 using cosine theta as a similarity measure. Numbers along bottom of dendrogram are same as those used in Table 5.1 to identify each individual specimen. Method by which these groupings are formed can be seen from Figure 5.3 by noting that similar objects lie on lines drawn through the origin, that is similarity is judged on constancy of proportions.

3. *Weighted-pair group average* – which is the same as for previous method except that the average is calculated differently. When a new object is added to a group the new average is taken as the mean of the new object and the old group average. This indicates that once objects are formed into a group the group is treated as if it were a single object. This seems to be one of the most generally used methods, possibly because the computation is somewhat easier and less time consuming.

There also are many similarity measures that have been proposed and among those in general use are:

1. *Euclidean distance* – simply the distance between two objects in n-dimensional space and should, of course, only be used if all the measures are on the same scale of measurement. The smaller the value the more similar the objects are. The euclidean distance between the ith and jth object, d_{ij}, is defined as:

$$d_{ij} = \sqrt{\sum_{k=1}^{n}(x_{ik} - x_{jk})^2} \tag{4}$$

2. *Correlation coefficient* – the cosine of the angle subtended by two objects at the mean point of all the data in n-dimensional space. The correlation coefficient between the ith and jth object, r_{ij}, is defined as:

$$r_{ij} = \frac{\sum_{k=1}^{n}(x_{ik}-\bar{x}_i)(x_{jk}-\bar{x}_j)}{\sqrt{\sum_{k=1}^{n}(x_{ik}-\bar{x}_i)^2 \sum_{k=1}^{n}(x_{jk}-\bar{x}_j)^2}} \tag{5}$$

3. *Cosine theta coefficient* – the cosine of the angle subtended by two objects at the origin of the n-dimensional space. It is useful for objects which vary alot in absolute size, but which maintain their shape or proportions, for example fossils. The nearer the cosine theta

coefficient is to 1 the more similar the two objects are. The cosine theta coefficient between the ith and jth object, $cosq_{ij}$, is defined as:

$$cosq_{ij} = \frac{\sum_{k=1}^{n} x_{ik}x_{ij}}{\sqrt{\sum_{k=1}^{n} x_{ik}^2 \sum_{k=1}^{n} x_{jk}^2}} \tag{6}$$

Similar to PCA, there is no right and wrong method, but an appropriate similarity measure must be chosen to suit each set of data. If in doubt try several methods and if fairly stable groupings occur then one can be reasonably sure that the answer may have some physical meaning, and not just be a quirk of the mathematics. As Marriott (1974) has pointed out, under certain circumstances cluster analysis can form clusters when none exist and fail to locate them when obvious ones exist.

For further details refer to the excellent discussions of the method in Kendall (1973), Hartigan (1975), and Gnanadesikan (1977),and refer to Le Maitre (1982) for geological examples.

5.5.1 Example 4

The *Fusispirifer* data from Table 5.1 has been used to illustrate cluster analysis using both Euclidean distance and cosine theta as similarity measures and the weighted-pair group average linkage method. The corresponding dendrograms are shown in Figures 5.6 and 5.7 respectively.

If one looks at the PCA plot of *Fusispirifer* data given in Figure 5.3, which shows the relationship of the individuals to each other, one can see how the groupings of Figure 5.6 have been obtained. For example, the left-most group in Figure 5.6 contains the four specimens 20, 22, 21, and 19, which can be seen to be clustered together in Figure 5.3 - they are all juveniles. Similarly, the next group in Figure 5.6 (17, 11, and 9) are also grouped loosely together in Figure 5.3 and so on.

Using, the cosine theta similarity measure, however, gives very different results as might be expected. For example, the right-most group in Figure 5.7 consists of specimens 17, 9 and 1 which, if one looks at Figure 5.3, are separated widely in space, but have *similar proportions* that is they lie on a line drawn through the origin. Similarly, the two juvenile specimens 20 and 19 are grouped with the submature specimens 13 and 12 because they all have similar proportions.

In this situation which method is to be preferred is entirely a paleontological decision in the hands of the user.

References

Blackith, R.E. and Reyment, R.A., 1971. *Multivariate morphometrics*: Academic Press, London. 412p.

Cattell, R.B., 1965, Factor analysis: an introduction to essentials: *Biometrics*, v. 21, p.190-215.

Francis, I., 1978, Factor analysis: fact or fabrication: *Math. Chronicle*, v.3, p.9-44.

Gnanadesikan, R., 1977, *Methods for statistical data analysis of multivariate observations*: John Wiley & Sons, New York, 311p.

Hartigan, J.A., 1975, Clustering algorithms: John Wiley & Sons, New York. 351p.

Jöreskog, K.G., Klovan, J.E., and Reyment, R.A., 1976, *Geological factor analysis*: Elsevier, Amsterdam, 178p.

Kendall, M.G., 1973, The basic problems of cluster analysis, *in* Cacoullos, T.ed., *Discriminant analysis and applications*: Academic Press, New York. p.179-191.

Le Maitre, R.W., 1982, *Numerical petrology*: Elsevier, Amsterdam. 281p.

Marriott, F.H.C., 1974, *The interpretation of multiple observations*: Academic Press, London, 117p.

Miesch, A.T., 1976, Q-mode factor analysis of compositional data: *Computers & Geosciences*, vol.1, p.147-159.

Spearman, C., 1904, General intelligence objectively determined and measured: *Am. Jour. Psychol.*, v.15, p.201-293.

CHAPTER 6

Use of Microcomputers in Building a Stream Sediment Database for Mineral Exploration

J.J. Durham
British Geological Survey
Keyworth, Nottingham U.K.

The British Geological Survey is undertaking a systematic regional geochemical survey of the United Kingdom landmass on behalf of the Department of Trade and Industry using stream sediments as the sampling medium. To date, samples have been collected from the whole of Scotland, northern England, and parts of central England and north Wales. The results are published in a series of geochemical atlases of Great Britain at a greater sample density, precision, and accuracy than any comparable survey and also are made available for use by universities, mining companies, and consultants. In this description, the emphasis is on the database building aspects of the survey, in particular the precautions taken to ensure that the data are valid, error-free, and self-consistent. Some methods of presenting the data for analysis and also some simple interpretive techniques are outlined. The aim is to give an overview of the British Geological Survey's project and present a generalized picture of how regional geochemical data may be used for mineral exploration.

6.1 REGIONAL GEOCHEMICAL SURVEYS

A regional-scale survey does not attempt to prepare a detailed map of a small area of geologic or economic interest. Rather, it attempts to cover a large area at an even sample density and to identify the broad geochemical patterns.

Regional geochemical surveys are a comparatively recent development, dating from the early 1950s, and only made possible by the development of rapid, low-cost, high-throughput analytical methods and computer-based information handling systems. In the United Kingdom the techniques were pioneered by Prof. J. S. Webb at Imperial College, London, whose work culminated in the publication of the Wolfson geochemical atlas of England and Wales (Webb and others, 1978). The Geochemical Survey Programme (GSP) of the British Geological Survey (BGS) is currently conducting a regional survey which aims to cover the UK landmass at a high sample density (1-2 per km^2), publishing the results both as conventional geochemical atlases and in computer-usable forms such as computer databases and on image processing systems.

6.1.1 Materials

For a regional survey, a material is required such that a single sample will be representative of the geochemistry of a large area. Some candidate materials are: rocks, soils, stream sediments, waters, and plants.

The optimum material to use in the particular area under study is best found by trial and error, that is an orientation survey must be made using different materials. For the GSP rocks, soils, and stream sediments were considered initially for the sampling medium.

Rocks are unsuitable for a regional-scale survey in the United Kingdom owing to their limited exposure at the surface and the occurrence of areas of deep weathering and alteration. There also is the problem of representing heterogeneous assemblages by statistically valid sampling models, and the difficulty of obtaining samples from faults and other structures, which often are areas of low topography infilled with deep overburden.

Soil sampling presents problems associated with the wide variation of soil types, the limited soil cover in upland areas, the wide variation in pH and Eh which affects the solubility and concentration of metals, and also the difficulty of ensuring consistent sampling of specific soil horizons by inexpert sampling teams. Both soil and rock samples convey information about a comparatively limited area, necessitating collection of large numbers of samples with consequent increase in the cost of the resulting geochemical map.

After rejecting rocks and soils, the GSP selected stream sediments for its sampling medium. Each stream sediment sample is considered to approximate to a composite sample of the weathering products of the rocks and overburden upstream, and hence to reflect the relative abundance of trace elements in bedrock of the catchment area. In the Scottish lowlands and parts of northern England, the reduced drainage density and presence of widespread contamination by agricultural and industrial waste has led to the introduction of low density soil sampling in parallel with the stream sediments.

6.2 DATA COLLECTION

For a comprehensive description of the sampling, analytical, and error control techniques used by the GSP, see the introduction to each geochemical atlas, for example the Great Glen geochemical atlas (BGS, 1987), and the references cited therein.

6.2.1 Sampling

A standardized sampling technique based on orientation studies carried out in the late 1960s is still used today with only minor modifications to improve sampling efficiency and reproducibility. Briefly, a minimum density of 1 sample per 1 km^2 is used, based principally on first-, second-, or occasionally third-order streams sampled 50-100 m above confluences. After removing the top 10-15 cm of sediment, the fraction smaller than 150 mm collected by 'wet' screening is taken, yielding a dry sample weight of approximately 50 g. At each sample site, a heavy mineral concentrate is prepared by panning about 5 kg of -2 mm sediment, and at most sites 30 ml water samples are collected for pH and conductivity measurements and further analysis at the field laboratory.

At each site, the detailed location is noted onto a preprinted 'field card' in the form of a free-text description based on a 1:50,000 topographic map and also the United Kingdom National Grid reference (to 100 m, omitting 100 km square). At the field base, the site location is plotted on a stable-base map and the grid reference refined to a full grid reference, accurate to ±10 m.

Also noted on the field card are codes to describe the catchment geology, any economic minerals seen in the pan, obvious contamination, and also free-text general comments. Because of the subjective and imprecise nature of this information, it is used mainly 'in house' to assist in interpretation of the data.

The data recorded in the field is transferred to a microcomputer database at the field headquarters by the field parties who collected the sample, immediately upon return to the field headquarters while the details are still fresh in the mind. Full checking and verifying of the field database is undertaken at the end of the field program. This approach is considered to be less error-prone than entering data many months later at BGS headquarters by data-entry staff.

6.2.2 Sample Processing and Analysis

Samples are freeze dried to disaggregate clay minerals, and following riffle splitting are subsampled and the particle size reduced by ball-milling to below 50 mm. Analysis is by simultaneous DC arc emission spectroscopy using a direct reading spectrometer. Data are collected for 26 elements and the results, corrected for inter-element and matrix effects, are stored on floppy disk and transferred later to a computer database.

The water samples collected at each site are analyzed at the field headquarters and a field laboratory for pH, conductivity, fluoride and bicarbonate ion concentration using conventional techniques (meters and titration methods). The results are entered into microcomputer databases. For bicarbonate ion analysis, a program written in dBASE is used to calculate the HCO_3^- ion concentration from titration results and to store the results directly in a database. Subsamples of the sediments and the waters are analyzed for uranium using the delayed neutron activation method.

6.2.3 Error control

Because the GSP is a long term program, extensive error control measures must be taken to ensure consistency of results and to this end various error control precautions have been incorporated into the sampling and analytical procedures.

To eliminate sample bias resulting from the evolution of different sampling procedures, the field parties, which consist mainly of students, are trained and supervised by professional members of staff and work in pairs, rotated daily. The daily sampling area selected for each team is an irregularly shaped area reflecting drainage rather than mapped geologic boundaries. Any variation in element levels related to sampling parties then can be identified by overlaying sampling areas over plots of the geochemical data.

One of the most important error control measures is based on allocating randomized sample numbers in the field and analyzing in numerical sequence. This measure is aimed principally at determining short term within-batch errors, for example instrument drift or contamination, which will show up as a clearly visible trend in the analytical values. If samples were collected and analyzed in strict numerical order, any trend of this nature could be due to either instrumental problems or real trends in the samples. Sampling in random order will tend to average out sample trends, and highlight variations due to instrumental problems. This is considered to have advantages over systems where samples are collected consecutively and analyzed in random order because error in sample preparation also is monitored. It also is preferred over systems where samples are assigned different field and analysis numbers, as this leads inevitably to introduction of errors.

The present analytical procedure has been in operation since 1975. To ensure long-term consistency of results, standards are inserted into the random numbering sequence at a frequency of 10 per 100 sample numbers, the numbers used being taken from a random number list.

Several standards derived from carefully homogenized bulk stream sediments of different compositions are used, and the results are plotted and monitored as analysis progresses. Any corrections or reanalyses needed to ensure the long-term consistency and precision of data are made to the appropriate batches of samples. Figure 6.1 shows a grayscale map of Bi illustrating the effect of a change of instrument background level for one atlas area relative to adjacent ones. Data derived from standards analyzed during the total analytical period for all the atlas areas enabled this to be corrected.

To monitor sampling and analytical precision, a method based on analysis of variance is used. At approximately 2% of sample sites, duplicate samples are taken a few meters away from the routine sample, and each of these is subsampled in the laboratory, giving a total of four samples from each site. The analytical data from these samples then can be used to determine within site (i.e. total error in sampling, sample preparation and analysis) and between site variance (i.e. geochemical variation). All the different types of sample are numbered in the normal random numbering system and the status (routine, standard, duplicate, or subsample) of a particular sample is not known by the analyst.

6.3 DATA PROCESSING

6.3.1 Database systems

The GSP stream sediment database is a large one, holding upwards of 60000 sample descriptions, each with locational information and analytical values for approximately 35 elements, and is growing at the rate of 3-4000 samples per year. It also is a comparatively old database, started in the mid-1970s using the software of that era. The database resides within the proprietary database management system (DBMS) 'Oracle' running on a

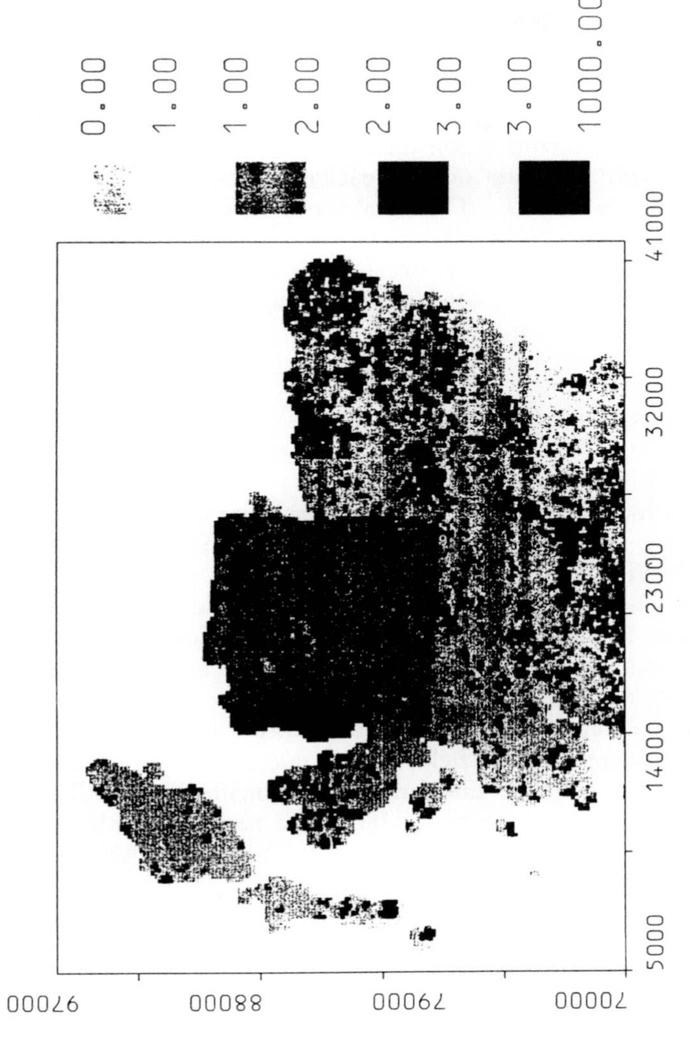

Figure 6.1 Grayscale plot showing effect of change in instrumental background level. Bismuth not analyzed over northernmost part of Scotland.

VAX computer at BGS headquarters. Microcomputers are used extensively for data collection and collation, the bulk of the preliminary data management being performed on IBM PC and Apple Macintosh computers, and the final 'clean' database then transferred to the mainframe Oracle database where it becomes the reference or 'archive' database.

The prime consideration for selecting microcomputer database systems for the GSP was ease of data exchange, particularly between IBM and Macintosh machines. The databases selected were dBASE III Plus for IBM and FoxBASE for Macintosh. When connected together over a local area network (LAN), these two DBMSs can use each other's database files directly, allowing easy combining of data.

6.3.2 Building the Database

Data has to be brought together from three different sources: the output from the analytical device; data from microcomputer databases; other data, hand coded.

The analytical data already is in an Oracle table maintained by the analysts. It must be written out to a text file, downloaded, and read into a microcomputer database where, after a few format changes, it will be ready for merging with the field data.

The field card data and water chemistry data, having been entered into microcomputers in the field, already are in dBASE III-compatible files, and need only to be transferred to the main dBASE III Plus database. This is done by means of the local area network to which all machines, IBM and Macintosh, are connected. Using a LAN means that any corrections to the field databases can be done by the field database managers, who maintain the reference database for their field party. When validating the data, dBASE III is used, accessing the (FoxBASE) field databases residing on the Macintoshes over the LAN, and only when the field databases are considered 'clean' are they transferred physically to the main (dBASE III Plus) database residing on an IBM PC.

Hard-copy data, for example lists of uranium analyses, must be entered into a microcomputer database, checked, and added to the main database.

The unique, arbitrary sample number assigned to each sample is used as the primary key in the DBMS, data being related between component databases and finally merged on this keyfield.

6.3.3 Error Checking

For a database to be of real use, it must be as error-free as possible: when using the GSP database for mineral exploration, there have been several situations where anomalous values of metalliferous elements have been determined to be caused by analytical standards remaining unnoticed in the database. This could cause needless and expensive fieldwork in following up such spurious anomalies. The task is to ensure that the analytical data, the field description, and ancillary information are all correct and associated with the correct sample.

In a database as large as the GSP, it is impractical to check every analytical result by hand against the hard-copy output from the analytical device, although this was done in the early days before machine-to-machine data transfer techniques were developed, and must still be done for any data entered by hand. Errors that do occur are often the result of aborted computer file transfers or format problems where the data stream has become out of step with the format describing it to the DBMS.

Using the facilities of a DBMS and a knowledge of the analytical results expected and the possible sources of errors, it is reasonably easy to devise a strategy to pick out most errors. Errors may often be detected by inspecting the extremes of values for the elements. For example, if nearly all copper values fall in the range 50-300 ppm and a small batch lies in the range 1,000-10,000 ppm, then there may be a format problem or a transposition with an element such as Ba; if an element is quoted as the oxide in percent it cannot have values over 100; pH values cannot be >14 and usually are in the range 4-8.5.

The location of each sample is normally taken from the field card information, although for some areas locations have been digitized from the sample sites plotted on stable-base topographic maps in the field. Locational errors can be checked by a knowledge of the sampling area: any locations falling outside this area, determined either by defining easting and northing limits or by plotting a site map and overlaying it on a topographic map, should be checked.

After a computer check for gross errors, a manual check of all sample locations is made. The site is located on the stable base map using the grid reference in the computer database and compared with the grid reference and free-text description of the site given on the field card. Any corrections or refinements are made, the field card corrected and amendments made to the computer database.

Errors in the flagging of samples (routine, standards, duplicates, etc.) potentially are disastrous and must be weeded out. Fortunately, the routine samples outnumber the others by an order of magnitude, so it is easy to produce lists of the nonroutine samples in the database and check them off against the master lists and the field cards. Some computer checking also is possible. For example, standards should have no associated locations and blank water samples should have no associated analytical data.

Finally, a proportion of the data must be thoroughly checked manually. The samples to be checked may be selected on the basis of random number generators, every nth sample or one or two per batch of results from the analytical device, but it is essential that a manual check be performed and that no errors are found. Only then can it be said that the data are ready for use.

The bulk of these data validation procedures are performed on the microcomputer database under dBASE III Plus. The size of the files at this point is around 6 Mbyte for data and index files, depending on the exact number of samples for the area being worked on. Most of the validation can be done from the dBASE III Plus command line, although some software has been written for such tasks as producing check plots, using the dBASE query language, often speed-enhanced by compiling using the commercial dBASE compiler 'Clipper'.

With such a large database being assembled, there is always the problem of version control. that is ensuring that the database being worked on currently is the latest version with all corrections to date incorporated in it. To this end, a central log is maintained where all additions and alterations to the main database are noted, together with the name of the person making the amendment and the file name of the resulting database. The component databases, for example the field location databases, are not merged into the main database until they are considered 'clean' and any amendments to these are always carried out on the computer holding that database. This is facilitated by working over a LAN; many problems previously arose when corrections were made to a working copy of a field database, not incorporated in the 'reference' field database and hence not incorporated in the main database when the reference copy was merged. At all stages, comprehensive backups must be made. Within the GSP, optical disks are used for database backups. These disks have a large capacity (200 Mbyte) which makes it possible to store backups at every stage of data processing, and are Write Once Read Many (WORM) devices rendering it impossible to delete or overwrite a backup file accidentally. After each change to the database, a check is made to ensure that the amendments were made as intended and that no accidental side effects were introduced. If all is correct this now becomes the current reference database, otherwise the previous version is reloaded and the amendments re-made.

When the database is finally considered clean, it is written out to a text file, uploaded to the VAX mainframe, read into an Oracle table and merged with the main GSP database. Any changes needed are now made to the Oracle database, which is the reference database, and copies are taken as required to keep the microcomputer versions in step.

6.3.4 Resources

The resources and manpower required for the generation of a database such as the GSP database should not be underestimated. As an example, consider the data processing effort of one of the two field parties.

Computing resources are at least one computer with printer and software, and a similar computer with the same software at BGS to act as a machine where any software problems can be reproduced and fixed, and also as a backup in the event of hardware failure in a field computer. Each computer has a hard disk, and sufficient floppy disks must be available to allow backups to be taken regularly. The final field database should exist as at least two identical copies on floppy disk as well as the hard disk version. Consumables such as printer paper and ribbons also must be provided.

Over a 10-week field season, collecting around 40 samples per day, approximately one hour of each evening will be spent entering the data into the field database. On return to BGS, complete checking and correction of the approximately 2000 samples collected, will take 2 people approximately 2 weeks to perform. The total manpower effort is thus 2 man-months per field party, per field season, merely to prepare the field data for merging with the analytical data. The final figure, taking into consideration both field parties and the time for generating the final, checked and complete database, will be of the order of 1-2 man years, depending on the size of the area being covered. This does not of course include the time taken for actual sampling and analysis, which increase the elapsed time to 2-3 years from commencement of sampling to release of data, with approximately 30 people being involved at one stage or another.

6.3.5 Data Quality

At this point, the database should be error-free and internally consistent. However, before being used, it is necessary to determine how much information the data actually convey. In mathematical terms: how much of the observed variation in geochemical values represents real variation in the sampled material and how much is an artifact of random variation in sampling and analysis. To check this, use is made of the subsamples and duplicates taken in the course of the sampling.

Using analysis of variance methods (BGS, 1987), the percentage of the total variance attributable to sampling and analysis is determined. The data set for this is approximately 1% of the number of routine sample sites, for example for the Argyll-Tiree geochemical atlas area containing approximately 10,000 samples, the data set consisted of 100 sites each having a subsample and a duplicate taken from a few meters away, itself subsampled, making 400 samples in all for the analysis. The results are given in terms of the sampling error, defined as the (within-site+within-sample variance)/total variance, expressed as a percentage.

Typical results are given in Table 6.1. For most elements, the between-set variance is significantly greater than the within-set variance, giving a low sampling error. This indicates that most of the variance occurring is due to geochemical variation and is therefore 'real' and meaningful. As would be expected, for those elements whose natural abundance levels are low, bordering on the detection limit, for example Be, Bi, the within-set variance is comparable with the between-set variance, giving a high sampling error. Care must be exercised when interpreting geochemical data based upon elements with such high 'false variance'.

Table 6.1 Analysis of variance for 100 samples and duplicates

Element	*Mean*	*Within-set variance*	*Between-set variance*	*Sampling error (%)*
Li	30.984	11.824	163.884	7.215
Be	1.218	0.636	1.363	46.670
B	29.363	73.678	888.742	8.290
MgO	3.894	0.616	11.406	5.400
K_2O	2.466	0.168	1.854	9.060
CaO	3.319	0.391	9.282	4.215
TiO_2	1.442	0.064	0.320	19.899
V	37.461	1621.755	9359.206	17.328
Cr	163.794	1693.025	35601.645	4.755
Mn	0.484	6.069	92.600	6.554
Fe_2O_3	9.347	2.081	11.270	18.466
Co	41.515	131.224	761.820	17.225
Ni	94.168	472.854	11372.103	4.158
Cu	38.348	301.660	1442.367	20.914
Zn	196.335	3139.183	16968.441	18.481
Rb	84.405	293.026	3615.411	8.105
Sr	279.657	2167.054	25556.865	8.479
Y	30.570	89.632	336.966	26.600
Zr	651.433	62230.279	231523.995	26.879
Sn	0.871	225.506	622.774	36.210
Ba	785.412	21437.160	166033.186	12.911
La	24.459	156.769	834.851	18.778
Pb	40.023	255.514	1262.841	20.233
Bi	1.405	4.944	9.112	54.261

Each set consists of two samples from each site, each sample subsampled and analyzed twice, that is 4 samples per set, 400 samples in total. Sampling error is (within-set/between-set)*100%.

6.4 USING THE DATABASE

The object of storing the stream sediment data within a database is to facilitate the use of the data for such purposes as mineral exploration, environmental geochemistry research, geochemical modeling, and so on. This involves querying the database and making selections on criteria such as anomalous values of various elements, usually over a particular geographical area or geologic unit. This subset of the data then may be used either within the database system for such processes as summary statistics, scatter plots, and formatted reports, or may be passed to other programs for more specialized processing. Transfer of data to other programs may be either direct, for example when passing dBASE III files to programs such as 'Statgraphics' which can read them directly, or via an intermediate file which the DBMS can write and the other program read. Such files are usually ASCII text files, either in a fixed format which must be defined to the receiving program, or as delimited files where the data fields are separated by special characters such as commas or tabs. As with all program-to-program file transfers, care must be taken to ensure that the correct fields have been written and have been read into the appropriate fields with the correct format.

Selection by geologic unit tends to be troublesome since the boundaries are rarely rectangular, and no general database has facilities for selection within irregular polygon limits. To overcome this, the GSP has written a dBASE III program (compiled with Clipper for speed) for selecting samples within an irregular polygon, but similar results can be obtained by selecting first an enclosing rectangle, then refining successively by deleting rectangular areas, finally plotting sample numbers on a base map and removing by hand the samples remaining outside the limits. For speed of processing, this subset is selected from a database consisting merely of sample numbers and locations, the relational features of dBASE III Plus being used later to relate this subset into the main database via the keyfield (sample number) to obtain the remaining data for each selected sample.

6.4.1 Data presentation

There are two main ways of simplifying and presenting the mass of data in a database of this magnitude meaningfully: statistical and graphical. The two methods are interdependent in that graphical methods must be based on a statistical analysis of the data, and graphical methods are often used to display the results of statistical analyses.

6.4.2 Statistical Methods

For mineral exploration, the areas of interest are those that show levels of an element that are enhanced to such an extent that they may indicate economically viable resources. Therefore, the first priority must be to perform a statistical analysis that will give the mean and variance of the overall distribution of each element and preferably a cumulative frequency distribution. A threshold can then be determined for each element above which values are considered anomalous. This criterion is subjective; typically being two or three standard deviations above the mean, or above the 95th or 99th percentile of the data.

Many statistical methods may be applied to the data but it must be borne in mind that, as is usual in the situation of 'real' data, the distribution is rarely normal, even after log transformation, and hence methods that require or imply normally distributed data should be used with caution. Methods that can be used to analyze the data include: histograms to display the distribution and indicate if two or more populations are present; principal component analysis; automated outlier rejection procedures, the high-level outliers being the points of interest. The methods may be applied to the entire dataset or to subsets defined on such criteria as geology, location or degree of contamination.

6.4.3 Graphical Methods

These methods rely on the fact that the human eye is perhaps the best tool for pattern detection. The aim is to present the data in the form of a map that will show the distribution of one or more element in such a way that anomalously high values can be identified readily and correlated with geologic and topographical features. The GSP utilizes various types of map for data presentation and interpretation, some generated from dBASE III files using GSP-written software, and some from commercial packages running on the BGS VAX and using data derived from Oracle files. Four types of map are shown in Figure 6.2.

6.4.4 Proportional Symbol Map (Fig. 6.2A)

Here, the symbol size is proportional to the data value. This is quite useful in that it can be overlaid onto a topographic or geologic base map and will not obscure detail to too great an extent. It also gives a broad indication of the geochemical pattern because the close spacing of larger symbols over an anomalous area makes them stand out relative to the background. The main problem is to select a symbol and size range that will show both the low and high extremes of the data: sizes that show up the lower values adequately will overlap excessively at upper levels rendering them illegible; reducing the size to show high values clearly reduces the lowest values to mere dots which may not plot on a conventional pen plotter. A variant using class intervals rather than a continuous range may be preferred in some situations.

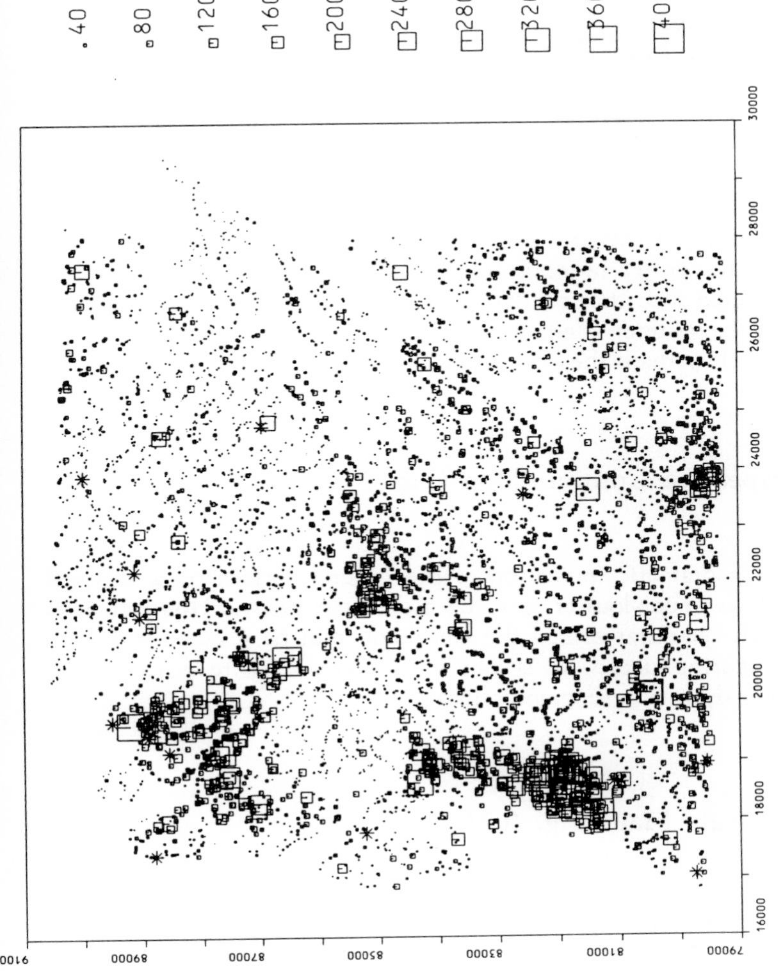

Figure 6.2A Example of proportional-sized symbol plot.

6.4.5 Posy-arm Map (Fig. 6.2B)

Similar to the proportional-symbol map, this easily can be overlaid onto base maps and is better from a quantitative standpoint because the arm length is easier to measure. The arm length scale must be selected with care especially if the data have a large distribution range. The angle of the arms may either be constant for all elements, in which case it should be selected such that overlap of the arms is reduced to a minimum, or they may be different, enabling maps for different elements to be overlaid. This is perhaps the best of the simple quantitative maps.

6.4.6 Perspective Contour Map (Fig. 6.2C)

This gives a dramatic indication of the general pattern on a regional scale and allows anomalous areas to be picked out easily. However, it is not particularly quantitative and the viewpoint must be selected with care to avoid obscuring data. It is not particularly easy to correlate an observed anomaly with a geographic location.

6.4.7 Grayscale Map (Fig. 6.2D)

By smoothing the data and binning into different class intervals, the broad patterns can be brought out. Maps of this kind may be available as lineprinter maps using different letters or combinations of letters to give the grayscale effect and thus are quick and cheap to produce. The cell size must be selected with care: if too large, the data become smoothed to such an extent as to average out small-scale variation;

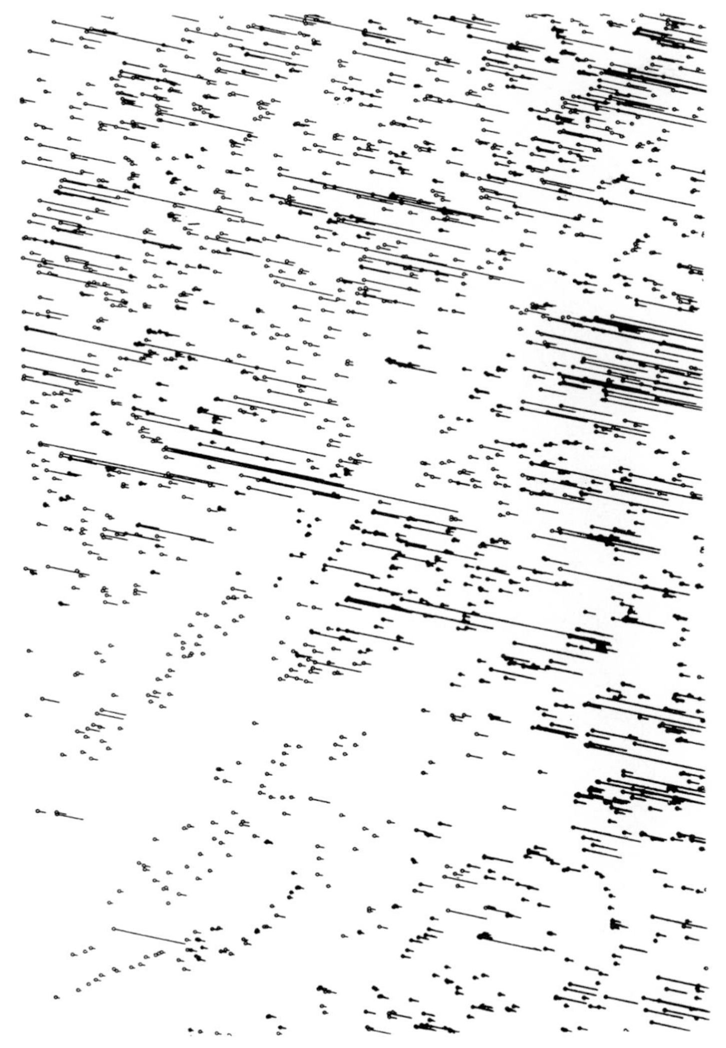

Figure 6.2B Example of posy-arm plot.

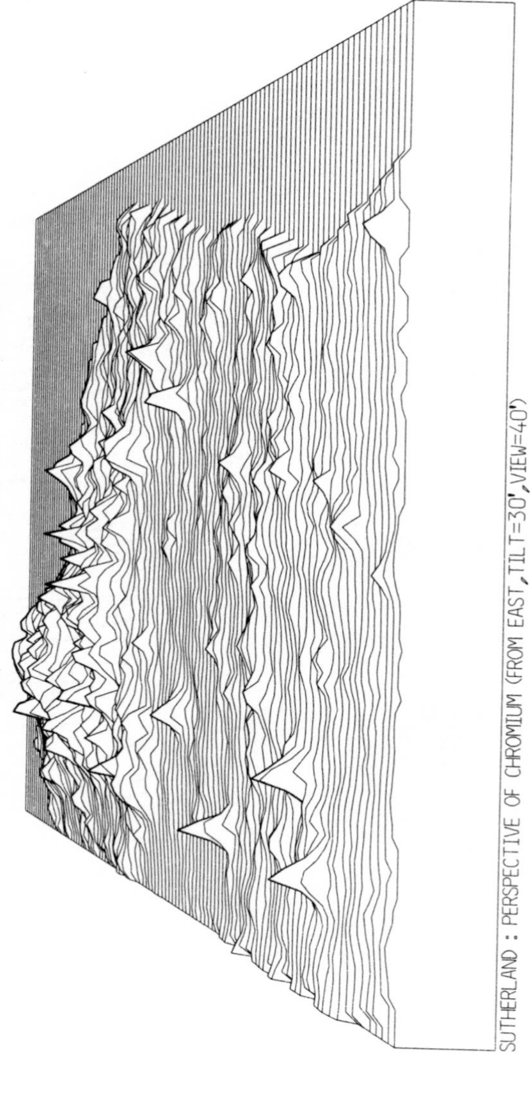

Figure 6.2C Example of perspective contour plot.

6.5.1 Direct detection

This method relies on the fact that a mineral deposit will usually show anomalously high values of the elements concerned. Mineralization involving Cu, Pb, Zn, Ba, and U may be detected from anomalously high values of those particular elements, either alone or in various combinations. For example, in Scotland, one Fe-Cu-Ni deposit shows up as anomalies of Cu, Pb, and Zn whereas another shows up as Pb, Zn, B, and other elements of basic association; uraniferous deposits almost always show U anomalies. In contrast, deposits involving Fe mineralization rarely show Fe anomalies, for example, Fe-Cu deposits tend to show anomalies of Cu, Zn, and Pb. In general, for the types of deposit located in Scotland, examination of maps for Cu, Pb, Zn, Ba, and U anomalies will show up areas of interest, especially those areas where anomalies of two or more of these elements coincide.

In addition to the elements of particular interest in the mineralization, other associated elements, known as pathfinders, are often found to have anomalous values. For example, with uranium mineralization in the sedimentary formations of Caithness and Orkney, Mn, B, and Mo usually are anomalous as are Pb, Zn, and Ba to a lesser degree. Gold exploration often uses Sb and As as pathfinders. Other useful pathfinder elements are Be, Sn, and the rare earths.

Problems occur in this type of interpretation of regional geochemical data because several of the elements that are important in mineral exploration are present also in artificial contaminants. If a stream is contaminated, a note is made on the field card giving the degree and type of contamination. Such contamination is typically due to farming (fertilizer bags, rusty wire fences, etc.), industry (effluents, empty containers for chemicals) or other human activity (tin cans, spent shotgun cartridges and pellets). These may alter the composition of the stream sediment by altering the biological or chemical composition of the stream and hence alter the trace element distribution. For example, many streams have large amounts of algae which concentrate many trace elements; others may have their chemistry altered by industrial discharges so as to cause certain elements to remain in solution when they would otherwise have been present in the sediment. Elements which often indicate contamination are Fe, Cu, and Sn.

Out of the mass of data in the database, a first scan through the data should concentrate on Pb, Zn, Ba, Cu, U, B, Be, Sn, and Mo, that is the economic minerals and pathfinders. Areas which show up anomalously high in at least two of these elements should be noted for further investigation. This further investigation should take the form of a scan through the raw data for the samples concerned, both analytical and descriptive (the field card information), firstly to check for errors and secondly to determine if there are any reasons to reject the sample as suspect. If the anomalies are judged to be real, and especially if they tend to reflect known mineralization trends as judged from studying the geologic base map, then they may be noted as areas worthy of further field study.

6.5.2 Deposit Modeling

This takes the procedure described previously several stages further. The anomaly associated with a known deposit is combined with all available information, geophysical, geologic and geochronological, and an attempt is made to produce a conceptual model for the deposit together with a geochemical, geologic and geophysical 'signature'. A computer search then can be made to locate similar areas, and hopefully a similar deposit. This approach, which relies on the availability of multidisciplinary databases is still being developed, but shows potential for locating deposits that have no surface geochemical expression.

6.6 EFFECTIVENESS OF GEOCHEMICAL PROSPECTING

The nature of a regional geochemical survey indicates that it is unlikely to detect a deposit directly. Rather it will act a a screening method to indicate areas of particular interest. It thus is difficult to assess directly the effectiveness or otherwise of geochemical prospecting. During the GSP survey, one base metal anomaly (Pb-Zn-Ba) when followed up resulted in the discovery of a barite deposit which is now being exploited commercially. What is unknown, and probably unknowable, is the extent to which the GSP has contributed to the directing of exploration funds away from barren areas.

Table 6.2 shows that the vast majority of mineral deposits in Scotland show geochemical anomalies, mainly of the metalliferous elements themselves. The few deposits that do not show geochemical anomalies are all vein-type deposits and occur either along the coast or in areas of poor drainage. In either situation there are too few samples available for reliable geochemical identification using stream sediments, although another sampling medium, for example soils, may well identify them. Probably the best indication of the value of the GSP database comes from the mining industry. Even in today's cost-conscious environment, the companies consider it worthwhile to spend several thousand pounds on purchasing regional geochemical data to help plan their own exploration programs.

Table 6.2 Mineral occurrences identified by geochemical anomalies

Type	*No.*	*No. with anomalies*	*Anomalies in metalliferous minerals (+ pathfinders)*	*Anomalies in pathfinders only*
Stratabound	7	7	5	2
Intrusions in Granites	17	17	17	-
Mineralization				
Old Red Sandstone	5	5	5	-
Vein deposits	115	110	75	-

6.7 ACKNOWLEDGEMENTS

The author would like to thank the other members of the Geochemical Survey Programme for their advice and comments when preparing this chapter, which is published with the permission of the Director, British Geological Survey.

dBASE III Plus, FoxBASE, Statgraphics and Oracle are trademarks of Ashton-Tate, Fox Software, STCS, and Oracle Corp., respectively. The mention of these products does not imply any endorsement of these products by the British Geological Survey.

References

British Geological Survey 1987, *Regional Geochemical Atlas*: Great Glen, British Geological Survey, Keyworth, Nottingham

Green, P.M. 1984, Digital image processing of integrated geochemical and geological information: *Jour. Geol. Soc. London*, v. 141, p.941-949.

Green, P.M. and Bide, P.J., 1986, Identification of mineral exploration targets by use of digital image analysis of integrated earth sciences information, *in Proc. Seventh Intern. Symposium on Prospecting in Areas of Glaciated Terrain, Kuipio, Finland*,: by Institute of Minerals and Metals, London, p. 49-60.

Plant, J.A., Smith, R.T., Stevenson, A.G., Forrest, M.D., and Hodgson, J.F., 1984, Regional geochemical mapping for mineral exploration in northern Scotland, *in Proc. Sixth Intern. Symposium on Prospecting in Areas of Glaciated Terrain*, Glasgow: Institute of Minerals and Metas., London, p. 103-120.

Plant, J.A., Forrest, M.D., Hodgson, J.F., Smith, R.T., and Stevenson, A.G., 1986, Regional geochemistry in the detection and modelling of mineral deposits, *in*: I. Thornton and R.J.Howarth, (eds.), *Applied geochemistry in the 1980s*, Graham & Trotman, London, p. 103-139.

Webb, J.S., Thornton, I., Thompson, M., Howarth, R.J., and Lowenstein, P.L. 1978, *The Wolfson geochemical atlas of England and Wales*, Oxford: Univ. Press, Oxford.

CHAPTER 7

Computer Exercises in Pattern Recognition in Exploration for Mineral Deposits related to Igneous Rocks

Glen R. Carter and Felix E. Mutschler
Dept. of Geology
Eastern Washington University
Cheney, Washington USA

7.1 INTRODUCTION

The mineral explorationist uses a variety of geologically oriented computer-hosted databases. These may include compilations of information on a single specimen or drillhole; results of local or regional mapping, geochemical, or geophysical surveys; production data from a mine, mining district, province, nation, or continent; and topical bibliographies on mineral deposits and related subjects. Such databases may be proprietary or publicly available compilations and may vary in size from a single record to hundreds of thousands of records. Today such databases may be used, in whole or part, on inexpensive personal microcomputers.

In this note we discuss the use of databases of "whole-rock" major-element oxide chemical analyses of igneous rocks to develop and test models of mineral deposits related to igneous rocks. Once such models are formulated, databases of igneous rock chemical data may be used as rapid access libraries to locate areas with a high exploration potential for the type of mineral deposit represented by the model. Although the technique and examples we discuss may seem simplistic, they essentially are first steps toward generating expertise that can be used in inference network artificial intelligence prospecting programs (see Campbell and others, 1982).

The basic premise of this technique is that certain types of mineral deposits show a close spatial and temporal association with certain chemically distinct types of igneous rocks. Examples include: (1) diamonds associated with kimberlites and lamproites; (2) chromite and/or platinum group element deposits associated with layered plutons derived from tholeiitic basalts; and (3) lode cassiterite deposits associated with two-mica, calc-alkaline granites. A second premise, that follows from the first, is that if a relation exists between a given type of mineral deposit and an identifiable type of source-host igneous rock, then areas containing the source-host rock are good places to prospect. In other words, recognition of such areas allows us to concentrate exploration efforts in limited high-potential areas, thus, hopefully, reducing the cost of discovery.

Exploration geologists always have accepted, at least intuitively, these premises. Our aim is to develop this prospecting technique in a quantitative fashion.

7.1.1 Exercise

A simple example may be helpful. A few decades ago a diamond exploration geologist might have tried to narrow the areas for field examination by doing a library search for all literature references to "diamonds", "kimberlites", "diatremes" ...etc. The method is valid, but in part depends on the assumption, often unwarranted, that all authors used the same criteria to define, say, "kimberlite". Even if all authors had used the same modal definition of kimberlite, such as that given by Clement and others (1984, p. 223-224)...

> "Kimberlite is a volatile-rich, potassic, ultrabasic, igneous rock which occurs as small volcanic pipes, dykes, and sills. It has a distinctively inequigranular texture resulting from the presence of macrocrysts set in a finer-grained matrix. This matrix contains, as prominent primary phenocrystal and/or groundmass constituents, olivine and several of the following minerals: phlogopite, carbonate (commonly calcite), serpentine, clinopyroxene (commonly diopside), monticellite, apatite, spinels, perovskite and ilmenite. The macrocrysts are anhedral, mantle-derived, ferromagnesian minerals which include olivine, phlogopite, picroilmenite, chromian spinel, magnesian garnet, clinopyroxene (commonly chromian diopside), and orthopyroxene (commonly enstatite). Olivine is extremely abundant relative to the other macrocrysts, all of which are not necessarily present. The macrocrysts and relatively early-formed matrix minerals are commonly altered by deuteric processes, mainly serpentinization and carbonatization. Kimberlite commonly contains inclusions of upper mantle-derived ultramafic rocks. Variable quantities of crustal xenoliths and xenocrysts may also be present. Kimberlite may contain diamond but only as a very rare constituent."

...the user of a database of whole-rock chemical analyses of igneous rocks such as IGBA, CLAIR, or PETROS is still at a loss until such terms as "volatile-rich","potassic", and "ultrabasic" are quantified in terms of the major-element oxides in the database. The LES.KIMB file is a small database of analyses of 42 kimberlite specimens from Lesotho extracted from PETROS. Let us try to develop a list of major-element attributes defining kimberlites so that we may screen other databases for kimberlites. Keep the guidelines in Table 7.1 in mind.

(1) Examination of LES.KIMB either visually, or by using the computer to construct histograms of various oxide values, suggests several possibilities. For example, 39 of the 42 analyses have SiO_2 contents between 20 and 40 wt%. We may consider this as a positive attribute and, if we wish, evaluate its statistical and petrogenetic significance.

Table 7.1 Suggestions for building screens for use with petrochemical databases to evaluate mineral potential

(1) Develop deposit model. Descriptive models of Cox and Singer (1986) and Ekstrand (1984) are good starting points.

(2) Build database of information from deposits of type you wish to screen for.

(3) Be sure that attributes you plan to use are represented in database. Often most obvious or desired variable is <u>not</u> available, simply because it is missing from database. For example, few whole-rock chemical analyses include low-level (ppb) gold determinations.

(4) Build your screen one attribute at a time.

(5) Screening is best accomplished starting with variable that allows largest subset of records to pass the screen.

(6) After each screening check resulting subset of accepted records to determine number of records which passed screen and how data in subset are distributed. Just as important is evaluation of data distribution in records rejected by screen.

(7) To tighten screen add one attribute at a time. Repeat until desired subset of records is obtained.

(8) Be sure that your screen is geologically reasonable.

(9) Knowing when to stop adding attributes to screen is as important as knowing what to screen for. The simpler the screening method, the better.

(10) Results of any analyses (economic or statistical) are no better than model or screening method. Do not hesitate to change either model or screen.

(11) Remember that this process only is one of many steps in evaluating mineral potential of area.

(12) Never forget that you will not find something that is not present in database.

A test run to select all analyses with 20-40 wt % SiO_2 from the 37,297 analyses in PETROS found 886 examples that fit the screen. Inspection shows many of these examples are clearly not kimberlites, suggesting that a screen based on SiO_2 alone is too open.

(2) Let us try to tighten the screen. Other attributes that inspection of the data might suggest include Al_2O_3 <10 wt % (42 examples) or <8 wt % (39 examples); MgO >12 wt % (41 examples); CO_2 ≥1 wt % (31 examples). A reading of Clement and others' (1984) definition suggests other possible attributes based on K_2O, K_2O/Na_2O, TiO_2, P_2O_5, etc.

(3) Try to develop your own two- or three-attribute screen for kimberlites and test it against file KIMB.TEST which contains several kimberlites. Do not forget to watch out for the pitfalls listed in Table 7.1.

(4) When you feel you have developed a satisfactory screen for kimberlites, you may wish to consider how else to use your database to locate diamond prospects. Two ideas which suggest themselves to us are:

(a) Locating analyses (and thus localities) for upper mantle-derived ultramafic rocks which, as Clement and others (1984) mention, are common inclusions in kimberlites. File KIMB.TEST contains 42 mantle inclusions from Lesotho kimberlite pipes (major group code LES.MANT). See if you can isolate some of them. An initial hint-try opening up your SiO_2 screen to accept values of 20 -50 wt %. A follow-up approach might involve constructing a screen based on the estimated lherzolite compositions given in Table 7.2.

(b) Diamonds have been discovered in "lamproites" in Western Australia.

Table 7.2. Estimated compositions of lherzolites (from Maaloe and Aoki, 1977)

	Spinel lherzolite	*Garnet lherzolite*	*Suggested primitive mantle*
SiO_2	44.20	44.99	44.71
Al_2O_3	2.05	1.40	2.46
FeO*	8.29	7.89	8.15
MgO	42.21	42.60	41.00
CaO	1.92	0.82	2.42
Na_2O	0.27	0.11	0.29
K_2O	0.06	0.04	0.09
TiO_2	0.13	0.06	0.16
P_2O_5	0.03		0.06
MnO	0.13	0.11	0.18
NiO	0.28	0.26	0.26
Cr_2O_3	0.44	0.32	0.42

Based on analyses recalculated to 100%, H_2O and CO_2 free.
FeO* = total iron as FeO.

A recent review by Mitchell (1985) includes the following attributes of lamproites:

(i) Lamproites are potassium- and magnesium-rich rocks of lamprophyric aspect.

(ii) They usually are leucite-bearing, but do not contain kalsilite, melilite, or nepheline.

(iii) They typically show the following composition: SiO_2 = 43-55 wt %; K_2O = 4-12 wt %; K_2O/Na_2O >5; CO_2 generally absent or low (compare this with kimberlites).

(iv) Usually they have peralkaline tendencies and show normative quartz, acmite, and hypersthene.

(v) Enrichment in Ti, REE, Rb, Sr, Ba, Zr coupled with high Ni, Cr, Co, and Sc contents, imparts a distinctive geochemical signature.

See if you can build a screen to isolate the lamproites in file KIMB.TEST.

Mutschler and others (1985) pointed out that "mafic ultrapotassic rocks" defined as having MgO >5.7 wt %, K_2O >6.0 wt %, TiO_2 >1.5 wt %, and P_2O_5 commonly >1.2 wt %, are not associated with precious metal (Au, Ag, Pt) deposits. Are lamproites mafic ultrapotassic rocks according to the above definition?

Finally, in your search for diamonds, do not forget other commodities and deposit types. For example, Mutschler and others (1978) suggested that the emplacement of kimberlites may be preceded by a sequence of tholeiitic and alkalic to carbonatite magmatic events-each of which may have associated mineral deposits.

7.2 SOME CASE HISTORIES

7.2.1 Granite Molybdenite Systems

By the early 1970s it was apparent that most of the world molybdenite production was a product of "porphyry" deposits where molybdenite was associated genetically with epizonal plugs and stocks. It was further recognized by many workers that these porphyry molybdenite deposits could be subdivided into two types with the attributes shown on Table 7.3. Economic analysis suggested that discovery of "Climax"- or "granite"-type molybdenite deposits could be profitable. In North America a variety of approaches to target definition and recognition of these deposits was used by explorationists. These included: (1) cataloging and examining reported occurrences of molybdenite mineralization; (2) detailed

Table 7.3 Classification of porphyry molybdenite deposits

Attribute	*Deposit type*	
	Granite molybdenite systems (Climax type)	*Granodiorite molybdenite systems (porphyry copper type)*
Source-host rock	High-silica granite and/or rhyolite	Granodiorite and/or quartz monzonite
Depth of emplacement tectonic setting	Epizonal (related in space and time to major calc-alkaline batholiths)	Epizonal to mesozonal (related in space and regional extensional stress)
Ore deposits	Stockwork Mo (Sn, W) Significant related base-metal deposits not usual	Stockwork Mo-Cu (Au, Ag) Significant related base metal skarn, vein, manto deposits are usual
	Mo >> Cu	Cu≥Mo (grade to porphyry copper systems without significant Mo)
Hydrothermal alteration assemblages	Propylitic Argillic Quartz-sericite ore Potassic ore Silicic zone	Propylitic Argillic Quartz-sericite ore Potassic zone

geological, geochemical, and geophysical studies of known deposits; and (3) theoretical and experimental evaluation of molybdenum transport and deposition in silicate melts and hydrothermal solutions.

Data from such studies allowed Mutschler and others (1981) to evaluate major-element oxide analyses of unaltered and altered granite and rhyolite host- and source-rocks associated with granite molybdenite systems in the western United States. They found that the unaltered source-host granites and rhyolites shared the following fingerprints:

(1) They are granites in the sense of Tuttle and Bowen (1958); that is, their norms show Ab+Or+Q ≥80%; and when Ab+Or+Q are normalized to 100%, they have Ab ≥20%, Or ≥20%, and Q ≥20% (see Fig. 7.1). All of these rocks are peraluminous (molecular Al_2O_3 > molecular Na_2O+K_2O) and therefore they contain normative corundum (C).

Such rocks are similar in some respects to common calc-alkali granites and alkali granites and rhyolites (see Table 7.4). To separate most of the unaltered rocks related to granite molybdenite deposits from similar granites and rhyolites not associated with molybdenite mineralization the following major-element fingerprint may be used:

SiO_2 ≥74.00 wt %, Na_2O ≤3.60 wt %, and K_2O ≥4.50 wt %.

(2) Granite molybdenite systems also show enrichment of some of the following elements: Be, Cs, F, Li, Mo, Nb, Rb, Sn, Th, U, and W when compared to the average values for low-calcium granites compiled by Turekian and Wedepohl (1961). However, attempts to define rigorous universally applicable trace-element chemical fingerprints for granite molybdenite systems have not been successful, probably because these rocks originate by partial fusion of crustal granulites which in different areas have different trace-element chemistries. Recognition that granite molybdenite system source rocks will be enriched in some of the trace elements listed previously is a valid exploration tool.

(3) Granite molybdenite deposits show a characteristic suite of hydrothermal alteration assemblages which can usually be related to the vertical level in the system, and to the MoS_2 orebody, as shown on Figure 7.2.

The chemical fingerprints summarized previously have been used successfully to recognize and evaluate granite molybdenite system prospects in the western conterminous United States. Limited experience

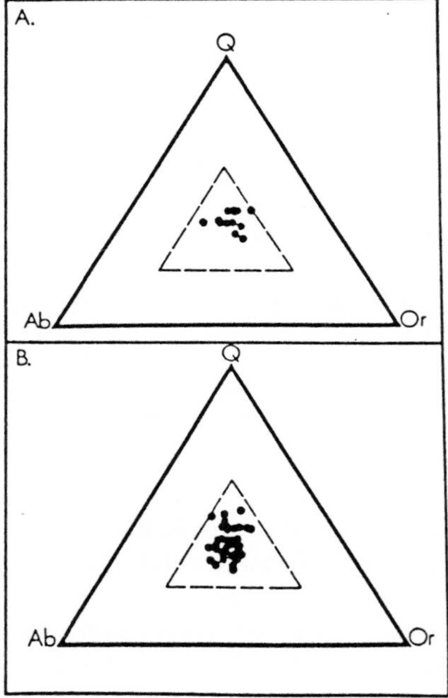

Figure 7.1 Ab-Or-Q diagrams for unaltered source rocks from granite molybdenite systems. On this and on Figure 10.3B-Ab=normative albite, Or=normative orthoclase, Q= normative quartz, and interior triangle defines granitic rocks in sense of Tuttle and Bowen (1958). A-13 analyses of source rocks from within or beneath ore bodies; B-50 analyses of source rocks spatially remote from ore. From Mutschler and others (1981, fig. 2).

Table 7.4 Mean composition of unaltered source rock plutons from granite molybdenite systems compared with average granite and rhyolite compositions

	1		*2*		*3*	*4*	*5*	*6*
	Mean	*SD*	*Mean*	*SD*	*Mean*	*Mean*	*Mean*	*Mean*
SiO_2	75.32	1.04	75.20	2.23	72.08	73.86	74.57	71.30
Al_2O_3	12.57	0.46	13.31	0.82	13.86	13.75	12.58	14.32
Fe_2O_3	0.83	0.36	0.70	0.83	0.86	0.78	1.30	1.21
FeO	0.55	0.26	0.42	0.38	1.67	1.13	1.02	1.64
MgO	0.46	0.23	0.24	0.23	0.52	0.26	0.11	0.71
CaO	0.73	0.32	0.66	0.45	1.33	0.72	0.61	1.84
Na_2O	3.27	0.38	3.64	0.48	3.08	3.51	4.13	3.68
K_2O	5.03	0.60	4.81	0.51	5.46	5.13	4.73	4.07
H_2O+	0.45	0.32	0.39	0.21	0.53	0.47	0.66	0.64
H_2O-	0.10	0.63	0.17	0.14				0.13
TiO_2	0.11	0.11	0.28	0.73	0.37	0.20	0.17	0.31
P_2O_5	0.09	0.44	0.05	0.06	0.18	0.14	0.07	0.12
MnO	0.06	0.05	0.04	0.03	0.06	0.05	0.05	0.05
CO_2	0.26	0.19	0.14	0.33				
F	0.21	0.21	0.11	0.11				
S	0.20	0.21	0.05	0.09				
Total	100.24		100.21		100.00	100.00	100.00	100.07

1 Ore related granite and rhyolite porphyries from molybdenite deposits (n=13); from Mutschler and others (1981, table 3).

2 Granite and rhyolite porphyries remote from ore at molybdenite deposits and prospects (n=50); from Mutschler and others (1981, table 3).

3 Average calc-alkali granite (n=72); from Nockolds (1954, Table I-I).

4 Average alkali granite (n=48); from Nockolds (1954, table I-III).

5 Average alkali rhyolite and obsidian (n=21); from Nockolds (1954, table I-IV).

6 Average granite-as named by authors (n=2485); from Le Maitre (1976, no. 8).

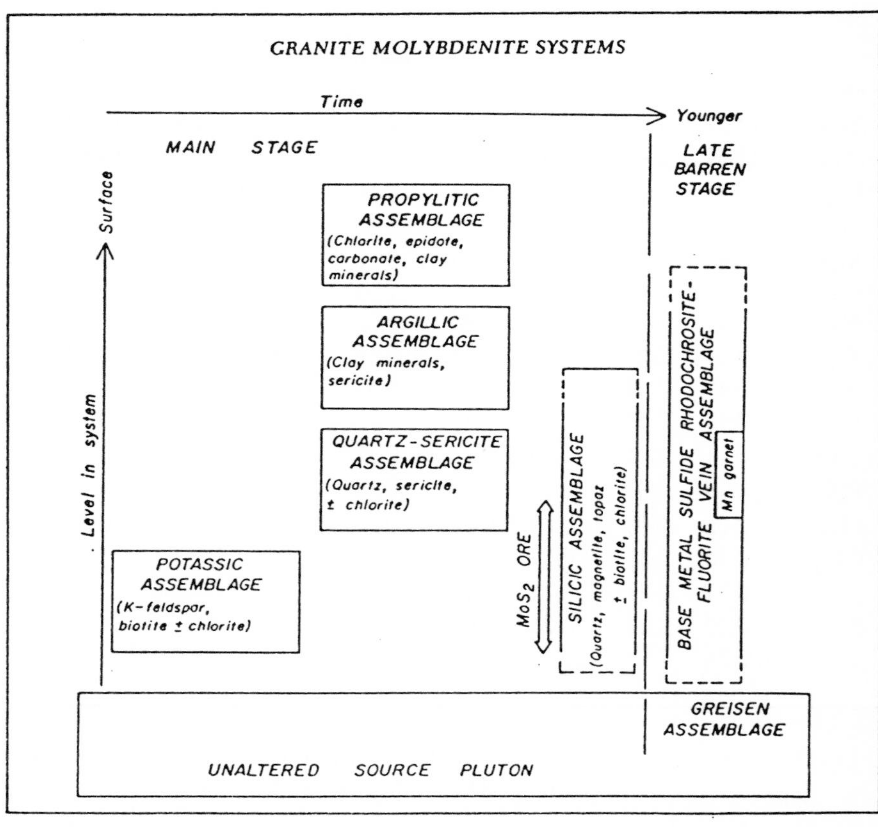

Figure 7.2 Diagrammatic representation of alteration assemblages in granite molybdenite systems. Index minerals used to recognize assemblage shown in parentheses. Clay minerals include kaolinite, dickite, montmorillonite, and pyrophyllite. From Mutschler and others (1981, fig. 3)

indicates that they also are applicable in eastern North America, Australia, and the North Atlantic Tertiary volcanic province.

7.2.1.1 *Exercise*

The file MOLY.US contains 103 analyses from seven granite molybdenite deposits in the western conterminous United States. You may wish to use the fingerprints given previously to separate these analyses into fresh and altered rocks, and to use the programs provided to plot data on Ab-Or-Q and A-K-F diagrams (see Fig. 7.3). MOLY.TEST contains 63 analyses from four granite molybdenite system deposits in the western Cordillera of British Columbia and Alaska. How do the fingerprints work with these data? Should you modify the fingerprints? If so, can you suggest a geological reason to explain the problem?

7.2.2 Uranium-Thorium Deposits Associated with Igneous Rocks

Various primary igneous and related hydrothermal uranium-thorium deposits associated with igneous rocks have been recognized. These include:

Ilimaussaq type (peralkaline, silica-undersaturated rocks) Primary magmatic deposits of U-Th-bearing silicates with potential co- or by-product Be, F, Nb, REE, Li, and Zr occur in the Precambrian Ilimaussaq alkaline complex of southwest Greenland (Sorensen and others, 1974). The unusual ore mineralogy and the close spatial and temporal relation of mineralization to strongly peralkaline lujarvites are results of strong silica-undersaturation and high alkalinity which restricted the separation of a hydrothermal phase from the silicate melt. High concentrations of incompatible elements thus were retained in the magma and ultimately were precipitated in rare phases including steenstrupine, monazite, thorite, rinkite, rhabdophanite, and eudialyte.

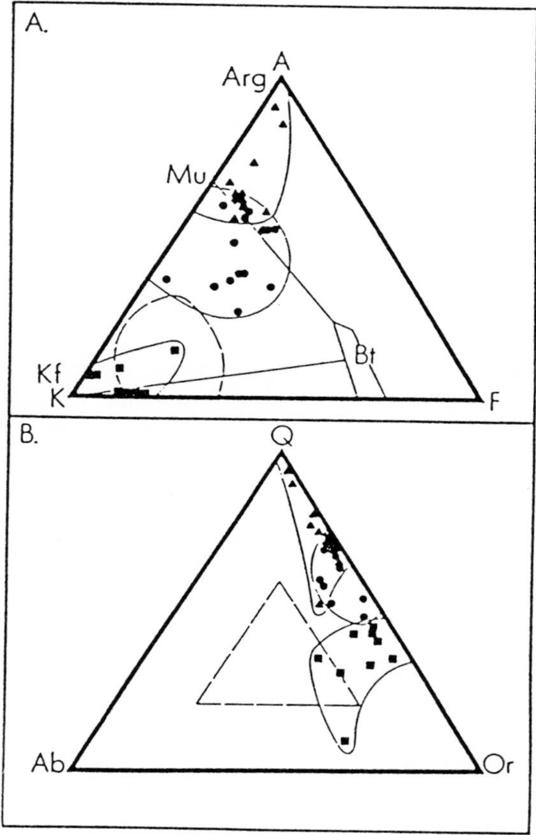

Figure 7.3 A- A-K-F, and B-Ab-Or-Q diagrams used to discriminate between alteration assemblages in granite molybdenite systems. Dashed lines enclose fresh source rock analyses used in Figure 7.1. Squares are samples from the potassic assemblage; circles are samples from the quartz-sericite assemblage; and triangles are samples from argillic assemblage. In (A), A=molecular Al_2O_3-molecular (Na_2O+K_2O+CaO); K=molecular K_2O; F=molecular (FeO+MgO+MnO); Arg=kaolinite, dickite, montmorillonite, and pyrophyllite; Bt=biotite; Kf=potassium feldspar; Mu=muscovite. From Mutschler and others (1981, fig. 4)

Younger granites of Nigeria type (peralkaline, silica-oversaturated rocks) Primary magmatic to hydrothermal uranium and thorium mineralization occurs chiefly as oxide minerals (uraninite, etc.) in disseminated grains or in veins associated with strongly oversaturated, highly alkaline to peralkaline plutons. Similar examples include the White Mountain magma series of New England, Bokan Mountain in southeastern Alaska, and the McDermitt caldera complex in Nevada and Oregon. Coeval, and cogenetic(?) Nb, F, Ta, Mo, W, Sn, and Zr mineralization may occur with these U-Th deposits.

Loon Lake type (peraluminous, low-calcium granites) Hydrothermal vein deposits, dominated mineralogically by U and Th oxides, characterize deposits associated with these specialized granites which occur often in batholiths. Other commodities associated with these systems include base-metals, Sn, W, Mo, Li, and F. The type example is the Loon Lake batholith, Washington. Other examples include the Cornwall granites of the United Kingdom and the "two-mica granites" of France.

Using major-element chemical analyses of igneous rocks from the examples cited previously Finn (1979) developed a series of chemical fingerprints to recognize rocks with a high potential for magmatic and magmatic-related hydrothermal U-Th deposits. These fingerprints include peralkalinity as measured by the agpaitic index ({mol Na+K}/mol Al); and positions on K-N-C (where K=wt % K_2O, N=wt % Na_2O, C=wt % CaO), A-M-F (where A=wt % Na_2O+K_2O, M=wt % MgO, F=wt % FeO+0.8998 Fe_2O_3) and C-Al-Ak (where C=mol Ca, Al=mol Al, Ak=mol Na+K) plots. These plots are shown on Figures 7.4A-C. All of these fingerprints are essentially differentiation indices in that they point out the most highly evolved members of different igneous lineages. Finn (1979) gives details showing how degree of alkalinity (measured by the agpaitic index) and the activity of silica in magmatic systems determine the U-Th mineralogy, ore grade, alteration assemblages, and spatial relation of ore bodies to the source plutons.

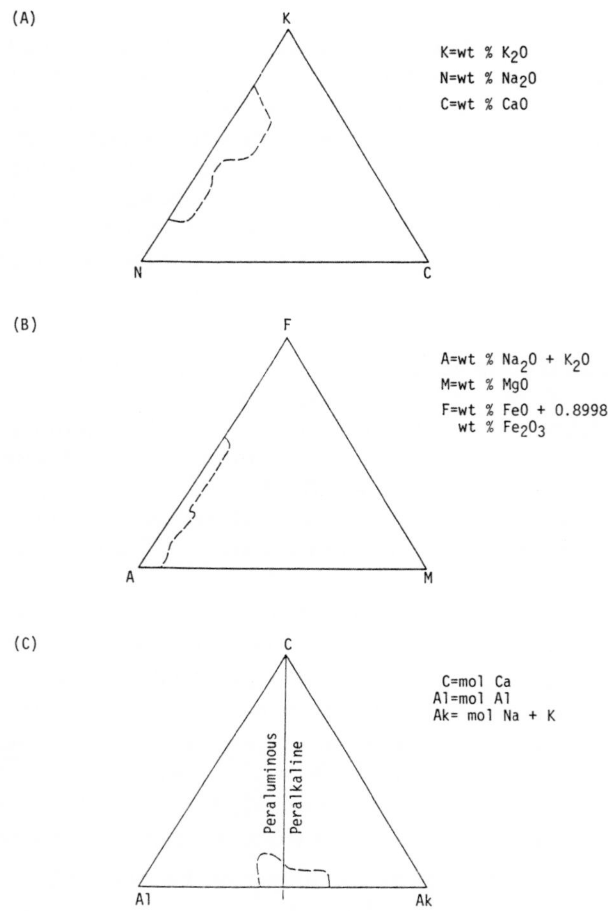

Figure 7.4 Triangular plots showing empirically defined composition fields (outlined) for igneous rocks with a high potential for economic U-Th (and other incompatible element) mineralization. (A) K-N-C plot; (B) A-M-F plot; (C) C-Al-Ak plot. Modified from Finn (1979, figs. 1-3)

7.2.2.1 *Exercise*

The file URAN) contains three data subsets: GGA (Ilimaussaq, Greenland), NIG (Younger granites of Nigeria), and BZM (Loon Lake, Washington) which correspond to the deposit types discussed previously.

Try K-N-C, A-M-F, and C-Al-Ak plots of each of the subsets. Histograms of the agpaitic index and X-Y plots of Na_2O+K_2O vs SiO_2 may be useful in pointing out differences between the deposit types in terms of source rock chemistry.

The Younger granites of Nigeria, especially in the Liruei Hills, include two suites of granites with distinctly different types of mineralization: (1) peralkaline granites (U-Th, Nb), and (2) biotite, or two-mica granites (Sn). How might you discriminate between these two suites?

Finally, you may wish to use PETROS data for evaluating other areas. We recommend computer trips to such diverse places as the Sierra Nevada batholiths (major group code SRN), the Bushveld Complex (major group code BUS), Afar (major group code AFR), and the Deccan Plateaus (major group code DEC).

7.2.3 Precious Metal Deposits related to Alkaline Rocks

About 12% of the total lode gold production of the U.S. and Canada has come from deposits which show a close genetic and spatial relation to alkaline igneous rocks. Included in these alkaline rock-related precious metal systems are a variety of deposit types (Mutschler and others, 1985). Two alkaline rock-related deposits (Kirkland Lake, Ontario and Cripple Creek, Colorado) have each produced more than 20 million troy ounces of gold, and eight other deposits either have production or reserves of at least 1 million ounces. This makes such systems attractive exploration targets.

To delineate areas with a high exploration potential for the discovery of alkaline rock-related precious metal deposits we suggest the following guides:

(1) Criteria to define potential source-host rocks.
   (a) They are alkaline in the sense of Macdonald and Katsura (1964); that is: wt % $Na_2O+K_2O > 0.3718$ (wt % SiO_2-14.5). See Figure 7.5.
   (b) They are members of coeval suites including: (a) alkali basalts and lamprophyres [basaltic rocks with normative olivine and nepheline (or leucite)], and (b) felsic syenites, trachytes, and/or phonolites with $Na_2O+K_2O > 10$ wt % and $MgO \leq 2$ wt %. The felsic rocks may be silica oversaturated, saturated, or undersaturated. Some typical suites are shown on Figure 10.5.
   (c) They show significant enrichments of Na, K, Rb, Th, U, F, Zr, Nb, and La and high Ba and Sr relative to Rb in Ba:Sr:Rb ratios as compared with average alkali basalts.

(2) Important pervasive alteration assemblages often may be recognized by the following whole-rock chemical criteria:
   (a) K-metasomatism- $K_2O > Na_2O$.
   (b) Redox alteration- $Fe_2O_3 > 1.5(FeO)$.
   (c) Carbonatic alteration- $CO_2 > 0.5$ wt %.
   (d) Phyllic alteration- appearance of normative corundum (sericite and/or clay minerals). Note that extensive hydrothermal alteration may have significant effects on the calculated norm (see Table 7.5).

7.2.3.1 *Exercise*

(1) Figure 7.5 shows total alkalis vs silica plots for three precious metal deposits associated with alkaline rocks in Colorado. Cripple Creek is

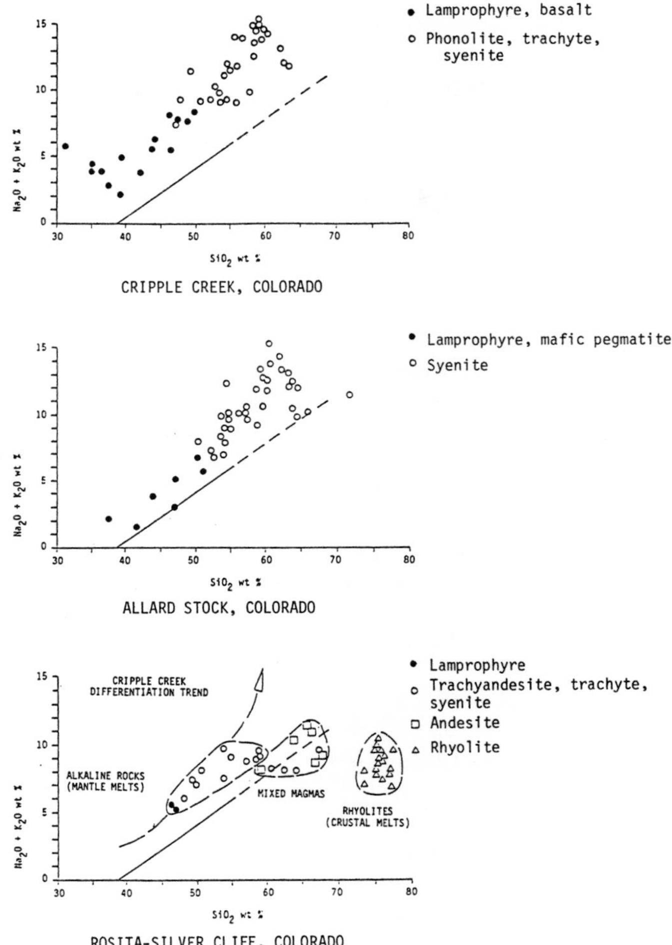

Figure 7.5 Wt % total alkalis-silica plots for source-host rocks from alkaline rock-related gold deposits. Diagonal line is Macdonald and Katsura's (1964) divider separating alkaline rocks (above line) from subalkaline rocks (below line).

Table 7.5 How hydrothermal alteration may change the CIPW norm --or how silica unsaturated rocks can become quartz normative

Carbonatic alteration

Hydrothermal CO_2 is added to rock. CaO used originally in forming normative silicates now is tied up in normative calcite (cc). This will produce excess SiO_2 which may result in calculation of normative quartz (Q). Reduction of normative anorthite (an) as result of carbonatic alteration will also produce extra Al_2O_3 which can occur as normative corundum (C).

Halogen metasomatism

Hydrothermal addition of F will result in calculation of normative fluorite (fr). This removes CaO which would have tied up SiO_2 in normative silicates, and may produce excess silica. Similarly addition of Cl will result in calculation of normative halite (hl). This removes Na_2O which would have tied up SiO_2 in normative silicates, leaving extra silica.

Phyllic alteration

Breakdown of silicates (particularly feldspars) to produce micas and clay minerals results in loss, via solution, of Na, K, and Ca, from rock. This produces, or increases the amount of excess silica. Excess Al_2O_3 produced by phyllic alteration will occur as normative corundum (C).

Pyritic alteration

Hydrothermal S added to rock results in calculation of normative pyrite (pr). This ties up FeO that otherwise might have gone into normative pyroxenes (di-hd, hy-en, hy-fs) and olivine (ol-fa). This will increase amount of uncommitted SiO_2 that may produce normative quartz.

Redox alteration

Oxidation of FeO to Fe_2O_3 will increase amount of normative magnetite (mt) and may produce normative hematite (hm). Thus, oxidation decreases amount of FeO available to combine with SiO_2 to form normative silicates, and may produce excess silica.

Silicification

Addition of hydrothermal SiO_2 can obviously increase 'silica saturation' of rock.

Solfataric (alunitic) alteration

The hydrothermal addition of sulfate (reported as SO_3) to rock will result in the calculation of thenardite (th) in norm. This reduces amount of sodium which would tie up silica in normative silicates and may yield excess SiO_2.

world-renowned for its gold telluride veins. The Allard stock in the La Plata Mountains hosts a porphyry copper deposit with significant concentrations of silver and platinum (see Werle and others, 1984). The Rosita district has produced rich gold telluride ores from breccia pipes related to paleo-geyser systems. What other type of ore deposit might be present in the Rosita-Silver Cliff district? How can you evaluate your suggestion?

(2) You are on your own now! You may wish to check the fingerprints suggested previously, and develop new ones using the data in ALKY, and then go on to solo prospecting in the libraries of CLAIR, PETROS, and IGBA. But do not spend too much time at your terminal. Remember-ORE DEPOSITS ARE DISCOVERED IN THE FIELD!

7.3 A SELECTION OF DATABASES AND PROGRAMS

ALKY (Griffin and others, 1985) is a fixed format database consisting of descriptive and major-element oxide and trace-element chemical data for 4208 specimens of alkaline and related igneous rocks from the western United States and Canada. Data were compiled from both published and unpublished sources and are divided into 86 groups representing petrologic provinces or geographic areas. ALKY has been used to develop and test a variety of ore deposit models. Preliminary descriptions of models for precious metal deposits related to alkaline rocks are included in Mutschler and others (1985).

Values for up to 145 variables may be stored for each specimen analyzed. Characteristics of the database are described in a 121 p. manual. Both database and manual are included on a 9 track 1600 BPI ASCII magnetic tape available from Eastern Washington University. ALKY was updated in 1984.

BIB.ALKY (Mutschler and others, 1986) is a computer-sortable bibliography containing 1123 citations for references on alkaline rocks and related mineral deposits in the North American Cordillera. Each reference has up to 15 keys which indicate the types of information in the reference. A FORTRAN program (BIBLIO) for sorting the bibliography by topic and/or geographic area is included. Both database and program are available in hardcopy; 9 track, 1600 BPI ASCII magnetic tape; or disk formats from Eastern Washington University. BIB.ALKY was updated in 1987.

BIB.MOLY (see Steigerwald and others, 1983b) is a computer-sortable bibliography containing 1187 citations for references on stockwork molybdenite deposits and related topics-with emphasis on the North American literature. Each reference has up to 15 keys which indicate the types of information in the reference. A FORTRAN program (BIBLIO) for sorting the bibliography by topic and/or geographic area is included. Both database and program are available in hardcopy; 9 track, 1600 BPI ASCII magnetic tape; or disk formats from Eastern Washington University. BIB.ALKY was updated in 1986.

BIBLIO is a FORTRAN program for selecting keyed records (literature references) from computer bibliographies in the formats used by BIB.ALKY and BIB.MOLY. Selected references are written to an output file which may be printed. A VAX-11 version of BIBLIO is included in hardcopy of BIB.MOLY (Steigerwald and others, 1983b) available from the U.S. Geological Survey, or on disk from Eastern Washington University.

CLAIR (Le Maitre, 1973) is a database consisting of descriptive and chemical data for 26,373 published analyses of igneous rocks from worldwide localities. Among other uses it provided the source data for Le Maitre's (1976) important contribution on the chemical variability of frequently occurring igneous rocks. It also was utilized in an early experiment in merging information from several large databases described by Granclaude and others (1980).

Information on the availability of CLAIR may be obtained from R. W. Le Maitre, Department of Geology, University of Melbourne, Parkville, 3052, Victoria, Australia.

FERROS (Carter, 1987a) is a fixed format database consisting of descriptive and major-element oxide and trace-element chemical data for 1320 specimens of banded iron formation and related rocks associated with gold deposits from worldwide occurrences. Data were compiled from published sources and are divided into 54 groups representing individual occurrences or groups of occurrences. FERROS was designed primarily to evaluate and test models for the recognition of gold deposits related to banded iron formations (Carter, 1987b).

Values for up to 54 variables may be stored for each specimen analyzed. Characteristics of the database are described in a 20 p. text file. Both database and text file are available on disks from Eastern Washington University. FERROS was updated in 1984.

GOLDY (Mihalasky and others, 1987) is a dBASE III Plus compilation of summary information on 112 North American mining camps that have reported production and/or reserves of more than 1 million troy ounces of gold. Included are data on location, geologic age, deposit type, production and reserves, ore grades and tonnages, and references. Information is current through the end of 1985. dBASE III Plus programs for data input and report generation are included with GOLDY which is available on disks from Eastern Washington University.

GPP (see Baker and others, 1985) is a user-friendly collection of fifteen integrated microcomputer programs designed to create, edit, and modify geochemical datafiles, perform a variety of petrologic calculations, and display and print graphical displays and data. It is especially useful as a tutorial system since it is keyed to A. R. McBirney's (1985) excellent text "Igneous petrology".

GPP, together with sample data files and illustrative exercises on disks, and an accompanying 44 p. hardcopy manual, is available from the Geology Department, University of Oregon, Eugene, OR 97403. Versions for Apple II+, IIe, or IBM PC are available.

GRANNY (Steigerwald and others, 1983a) is a fixed format database consisting of descriptive and major-element oxide and trace-element chemical data for 507 specimens of Laramide and younger high-silica rhyolites and granites from Colorado and north-central New Mexico. Data were compiled from both published and unpublished sources and are divided into fifteen groups representing geographic areas. GRANNY has been used to evaluate "chemical fingerprints" for the recognition of high-silica rhyolites and granites with a high exploration potential for the discovery of Mo (and other lithophile element) deposits (Mutschler and others, 1981).

Values for up to 106 variables may be stored for each specimen analyzed. Characteristics of the database are described in a 52 p. text file. Both database and text file are available in hardcopy from the U. S. Geological Survey (Steigerwald and others, 1983a); or on 9 track 1600 BPI ASCII magnetic tape or on disks from Eastern Washington University. GRANNY was compiled in 1983 and has not been updated since that time.

IGBA is a database containing published descriptive, major-element oxide, and trace-element chemical data for more than 15,000 igneous rock specimens worldwide. It is continually expanding its coverage as data are added by petrologists participating in Project 163 of the International Geological Correlation Program (see Chayes and Mutschler, 1978). For additional information on IGBA contact Felix Chayes, National Museum of Natural History, Smithsonian Institution, Mail Stop #129, Washington, D.C. 20560, U.S.A.

IGBA is available on magnetic tape from World Data Center-A for Solid Earth Geophysics, Code E/GCI, Boulder, CO 80303, U.S.A.

KEYBAM (see Barr and others, 1977) is a user-friendly system of interactive COBOL and FORTRAN programs for accessing and operating on information in database MARLA. KEYBAM includes routines for creating subfiles from MARLA using any or all of the MARLA variables, printing subfiles, plotting triangular diagrams (TRIPLT programs), and linking subfiles to the SPSS statistical package (Nie and others, 1975) which can produce a large variety of statistical analyses and graphical displays. Subfiles created from database MARLA by KEYBAM may be reformatted as dBASE III Plus records using a C language (standards of Kernighan and Ritchie, 1978) program. A user's guide to KEYBAM is included in a text file KEYBAM.INSTRUCT.

A version of KEYBAM which runs on the VAX-11/780 computer system is available from Eastern Washington University on 9 track, 1600 BPI, ASCII character set magnetic tape.

MARLA is an enhanced version of the PETROS database designed for use with the KEYBAM sorting programs. It is a fixed format database of descriptive and major element chemical data for 37,297 rock specimens of igneous rocks from worldwide locations. Included for each specimen analyzed are the 32 variables from PETROS and an additional 86 computed variables including: (1) oxide values recomputed to total 100% volatile free; (2) normative minerals in weight percent and cation equivalents; (3) differentiation indices; (4) agpaitic index; and (5) calculated rock name according to the scheme of Irvine and Baragar (1971). MARLA and KEYBAM are available on magnetic tape (9 track, 1600 BPI, ASCII) from Eastern Washington University.

PETROS (see Mutschler and others, 1976) is a fixed format database consisting of descriptive and major-element chemical data for more than 37,000 specimens of igneous rocks from worldwide locations. Data were compiled from both published and unpublished sources and are divided into 246 groups representing petrologic provinces or geographic areas. PETROS was designed for research and teaching and has been acquired by more than 200 institutions worldwide. It has been used to address and test a wide range of petrologic, tectonic, and economic geology problems.

Values for up to 32 variables may be stored for each specimen analyzed. Characteristics of the database are described in a 214 p. manual (MARTHA). Both database and manual are included on a 9 track 1600 BPI ASCII magnetic tape available from World Data Center-A, NOAA, Code E/GC1, Boulder, CO 80303, U.S.A. PETROS was updated in 1983.

References

Baker, B. H., McBirney, A. R., and Geist, D. J., 1985, GPP--A program package for creating and using geochemical data files: Eugene, Oregon, Department of Geology, University of Oregon, 44 *p.*, 2 computer disks.

Barr, D. L., Mutschler, F. E., and Lavin, O. P., 1977, KEYBAM--A system of interactive computer programs for use with the PETROS petrochemical data bank: *Computers & Geosciences, v. 3, p. 489-496.*

Campbell, A. N., Hollister, V. F., Duda, R. O., and Hart, P. E., 1982, Recognition of a hidden mineral deposit by an artificial intelligence program: *Science, v. 217, p. 927-929.*

Carter, G. R., 1987a, FERROS--A database of chemical analyses of banded iron formations and related rocks: Cheney, Washington, Eastern Washington University (magnetic tape).

Carter, G. R., 1987b, FERROS--A computer hosted database for geochemical analysis of rocks related to auriferous iron formations: Cheney, Washington, Eastern Washington University, M.S. thesis, *150 p.*

Chayes, Felix, and Mutschler, Felix, 1978, International Geological Correlation Program Project 163--IGBA--International database for igneous petrology--An invitation to participate: *Geology, v.* 6, *p. 543-546.*

Clement, C. R., Skinner, E. M. W., and Scott Smith, B. H., 1984, Kimberlite redefined: *Journal of Geology, v. 92, p. 223-228.*

Cox, D. P., and Singer, D. A. (editors), 1986, Mineral deposit models: *U.S. Geological Survey Bulletin 1693, 379 p.*

Ekstrand, O. R. (editor), 1984, Canadian mineral deposit types: A geological synopsis: Canada Geological Survey Economic Geology Report 36, 86 *p.*

Finn, D. D., 1979, Prospecting for magmatic and hydrothermal deposits of uranium and associated elements by computer evaluation of the chemistry and petrogenesis of igneous source rocks: Cheney, Washington, *Eastern Washington University, M.S. thesis, 336 p.*

Granclaude, Ph., Boudette, E., Le Maitre, R. W., and Mutschler, F. E., 1980, Grouped procession of the GUF, RASS, CLAIR, and PETROS files for chemical comparison of two-mica granites of various ages and locations: Sciences de la Terre, Serie Informatique Geologique, no. 13, *p. 95-109.*

Griffin, M. E., Mutschler, F. E., and Stevens, D. S., 1985, ALKY--A data bank of chemical analyses of alkaline and related igneous rocks from western North America: Cheney, Washington, Eastern Washington University (magnetic tape).

Irvine, T. N., and Baragar, W. R. A., 1971, A guide to the chemical classification of the common volcanic rocks: *Canadian Journal of Earth Sciences, v. 18, p. 523-548.*

Kernighan, B. W., and Ritchie, D. M., 1978, The C programming language: Englewood Cliffs, New Jersey, Prentice-Hall, 228 *p*.

Le Maitre, R. W., 1973, Experiences with CLAIR: *Chemical Geology, v. 12, p. 301-308.*

Le Maitre, R. W., 1976, The chemical variability of some common igneous rocks: *Journal of Petrology, v. 17, p. 589-637.*

Maaloe, S., and Aoki, K.-I., 1977, The major element composition of the upper mantle estimated from the composition of lherzolites: Contributions to Mineralogy and Petrology, v. 63, *p. 161-173.*

Macdonald, G. A., and Katsura, T., 1964, Chemical composition of Hawaiian lavas: *Journal of Petrology, v. 5, p. 82-133.*

McBirney, A. R., 1985, Igneous petrology: San Francisco, Freeman and Cooper, *509 p.*

Mihalasky, M. J., Mutschler, F. E., Etienne, J. E., and Gordon, T. L., 1987, GOLDY--A geologic and economic database for giant lode gold camps of North America: Cheney, Washington, Eastern Washington University (magnetic disks).

Mitchell, R. H., 1985, A review of the mineralogy of lamproites: *Geological Society of South Africa Transactions, v. 88, p. 411-437.*

Mutschler, F. E., Finn, D. D., and Ludington, Steve, 1978, Magmatism and related ore deposits of extensile continental environments--Computer exercises in petrochemical pattern recognition and prediction: Los Alamos, New Mexico, Los Alamos Scientific Laboratory Conference Proceedings LA-7487-C, *p. 64-65.*

Mutschler, F. E., Griffin, M. E., Stevens, D. S., and Shannon, S. S., Jr., 1985, Precious metal deposits related to alkaline rocks in the North

American Cordillera--An interpretive review: *Geological Society of South Africa Transactions, v. 88, p. 355-377.*

Mutschler, F. E., Griffin, M. E., Stevens, D. S., and Ludington, Steve, 1986, A bibliography of alkaline rocks and related mineral deposits, North American Cordillera: Cheney, Washington, Eastern Washington University (magnetic tape).

Mutschler, F. E., Rougon, D. J., and Lavin, O. P., 1976, PETROS--A data bank of major-element chemical analyses of igneous rocks for research and teaching: *Computers & Geosciences, v. 2, p. 51-57.*

Mutschler, F. E., Wright, E. G., Ludington, Steve, and Abbott, J. T., 1981, Granite molybdenite systems: *Economic Geology, v. 76, p. 874-897.*

Nie, N. H., and others, 1975, SPSS--Statistical package for the social sciences (2nd ed.): New York, McGraw-Hill Book Co., *657 p.*

Nockolds, S. R., 1954, Average chemical compositions of some igneous rocks: *Geological Society of America Bulletin, v. 65, p. 1007-1032.*

Sorensen, H., Rose-Hansen, J., Nielsen, B. L., Lovborg, L., Sorensen, E., and Lundgaard, T., 1974, The uranium deposit at Kvanefjeld, the Ilimaussaq intrusion, South Greenland: *Greenland Geological Survey Report 60, 54 p.*

Steigerwald, C. H., Mutschler, F. E., and Ludington, Steve, 1983a, GRANNY, a data bank of chemical analyses of Laramide and younger high-silica rhyolites and granites from Colorado and north-central New Mexico: *U.S. Geological Survey Open File Report OF83-0516, 562 p.*

Steigerwald, C. H., Mutschler, F. E., and Ludington, Steve, 1983b, A bibliography of stockwork molybdenite deposits and related topics (with an emphasis on the North American literature): *U.S. Geological Survey Open File Report OF83-0382, 112 p.*

Turekian, K. K., and Wedepohl, K. H., 1961, Distribution of the elements in some major units of the earth's crust: *Geological Society of America Bulletin, v. 72, p. 175-192.*

Tuttle, O. F., and Bowen, N. L., 1958, The origin of granite in the light of experimental studies in the system $NaAlSi_3O_8$-$KAlSi_3O_8$-SiO_2-H_2O: *Geological Society of America Memoir 74, 153 p.*

Werle, J. L., Ikramuddin, M., and Mutschler, F. E., 1984, Allard stock, La Plata Mountains, Colorado--an alkaline rock-hosted porphyry copper-precious metal system: *Canadian Journal of Earth Sciences, v. 21, p. 630-641.*

Index